THE ORIGINS SOLUTION

An Answer In The
Creation-Evolution Debate

DICK FISCHER

Fairway Press
Lima, Ohio

THE ORIGINS SOLUTION

FIRST EDITION
Copyright © 1996 by
Dick Fischer

Scripture quotations are from the *King James Version of the Bible*, in the public domain.

––––––––––––

Library of Congress Catalog Card Number: 95-60966

––––––––––––

ISBN 1-55673-188-4

PRINTED IN U.S.A.

"Seek the truth; come whence it may,
lead where it will, cost what it may."

The Reverend Dr. William Sparrow, 1801 - 1874
Professor of Theology and Christian Evidencer
Virginia Theological Seminary

TABLE OF CONTENTS

INTRODUCTION

We humans stand at the pinnacle of God's creation. Among the more curious of His creatures, we question and wonder aloud. We wrestle with the nature of things, and ponder the truth of matters. What do we do when the tab that someone labeled "A" looks like it fits better in slot B? What action should we take when those who seem to have the necessary credentials give insufficient answers to questions we think are important?

We are encouraged to be seekers of the truth, yet we are expected to have faith and believe. So what are we supposed to do when, in seeking the truth, we discover evidence that weighs against time-honored beliefs?

Do we pray for wisdom and discernment? Do we trust others who say they know the answers? Should we stay "in the Word," and keep focused narrowly on the biblical text, leaving tough questions for philosophers to handle? Perhaps there are things we are not supposed to know this side of heaven. Maybe we should take the attitude that we have the truth, don't confuse us with facts.

I rather suspect that you are not content with pat answers, or untidy loose ends, and neither am I. The reason I feel that can be said about you is that the title of this book, The Origins Solution, has aroused your curiosity enough for you to pick this book up and read at least this far. Others saw the title and passed it by without being the least bit inquisitive.

They have read other books that promised answers perhaps but failed to deliver the goods. I have read them; maybe you have too. They encounter the knot, and tell us they will untie it. They look at it, walk all around it, measure it, expound on it and in the end, cut it instead.

This book was written not just to ask questions, but to furnish answers long overdue. That does not mean everything is known about the origins of life or the origins of man, but that should not deter us from weighing the evidence we have and drawing sound conclusions based on the available data.

Can science and Scripture ever be reconciled? Probably not, but peaceful coexistence is not out of the question. That much can be said for both Scripture and science deal with truth. Science invokes a process that shuns miracles, however, while the Bible abounds in them. Miracles defy testing, and testing is part of science. On this score, religion and science will always remain at odds. Where the Bible is history, though, modern archaeology has only confirmed its accuracy.

In Scripture, we are shown the truth and told the truth, whereas in science we search for truth, or at least a plausible explanation, using proven methods. So it should come as no surprise when we find agreement, but when there seems to be discord, we may need fresh answers.

There was a time when many believed the earth was flat. Augustine considered that from his understanding of the biblical text. When we learned the earth was a globe, people did not toss the Bible. Some interpretations were adjusted, but we found Scripture would accommodate a round earth as cordially as a flat one.

Another brouhaha concerned the geocentric earth view with the earth at the center of the universe, versus the modern concept that the earth rotates on its axis and revolves around the sun. We can thank Copernicus and Galileo for putting us on the right track. What is notable is that the Bible had not warned its readers about this revelation of science to come.

Our forefathers had been allowed to hold erroneous convictions in the realm of earth science without the Bible trying to thrust them into the near modern age with a jargon that could not have been understood at the time. On the other hand, the inspired Book is not trapped by passé prose on a geocentric earth, or a flat one either.

These brief examples are cited to illustrate that we have a living and timeless Bible. In addition to the other remarkable claims we could make about Scripture, we can add one more. God's written Word can adapt to whatever we learn, whenever we learn it.

The reason for versatility is understandable. The Bible had to survive through diverse periods of history and be as relevant to the monks of the Middle Ages as it will be to future spacemen.

Space travelers and Middle Age monks do not share the same world view, but all can benefit from the same eternal message from a changeless God.

God's revealed Word in the Bible is called special revelation. Today we have numerous translations. We are awash in commentaries. If we can cut through the clamor, we can learn the specific things of God from His Word we could not know from observing nature alone. Illumination or understanding may be aided by the empowerment of the Holy Spirit.

General revelation is what we observe in nature. Anyone can see it, examine it, taste it, feel it, experience it, or test it. Available for all to scrutinize, the natural world around us should give us at least an inkling of the character of our Creator.

Through science, we try to unravel the mysteries of His general revelation. We have nothing to fear in seeking the truth in nature. Our Creator allows us to make our own observations, and honest reporting on our part is all that is expected.

At this juncture, no scientist has presented even a token amount of credible evidence that a universe can spring into existence unassisted; or that life can come about spontaneously from a chemical soup; or that a DNA molecule, the genetic code of life, can become increasingly complex through time entirely by accident, even if some try to persuade us in those directions.

Special revelation and general revelation should match up since God is the author of both the Book of Words, and the book of works. If the revelations of the Bible appear to contradict the revelations of science, it is due to error that has crept in someplace.

A large part of the unnecessary tension that seems to exist between science and the Bible today has been caused by outdated interpretation. Some traditional Bible "truths" being taught from pulpits and in seminaries today are stuck hopelessly in the 17th century, whereas our society is careening rapidly into the 21st century. It is past time to free up our living Bible to be completely relevant to coming generations.

We are not trapped on a flat, geocentric earth by an inflexible, outdated book. The inspired text is adaptable whether we are or not. For example, we live on a planet scientists date to about 4.6

billion years, and the Bible will concord with that. Notwithstanding, there are some who reckon our habitat was created instantly no earlier than 10,000 years ago. That's a whopping difference, but a living Bible is ready to move into the modern era with those who believe in a young earth the minute they pick up a geology book.

Biologists believe the fossil record expresses a sequence of life forms that evolved through time. Meanwhile, some Bible expositors have ventured that a world-wide catastrophic flood sequenced the dead animals, burying the simplest creatures first, thereby leaving a contrived data trail with increasing complexity merely a happenstance caused by some mystical sorting action.

Archaeologists have traced hominids back beyond four million years. We may not know who the first anthropoid to walk on two legs was, but with a high degree of certainty we can rule out Adam of Genesis. Still, some insist Adam was the first mammalian biped on earth. So who is right? What does the Bible say? We will examine these controversies and see they can finally be put to rest. Scripture is not mired in the tar pits of obsolete dogma.

Let me stress one other point. We can see predictable differences in Bible interpretation depending on whether that interpreter takes a liberal or a conservative stance. This book seeks to bridge those opposing viewpoints, yet affirms that a literal reading of the first eleven chapters of Genesis, when shorn of mistakes in translation and interpretation, harmonizes remarkably with ancient history and the latest findings of modern science. Attempts to reconcile Bible and science by categorizing difficult Old Testament passages as poetry, allegory, mythology, or literary fiction are entirely unjustified.

This book is about resolving centuries old conflicts. The solutions, though, are concentrated in the second half of the book. A lot of spade work went into the first half to lay a foundation so that it can be seen easily why the tendered solutions clearly outdistance the competition. Hopefully, no one will drop out midway.

So I have done my best to make this enjoyable. I hope you will find this book as satisfying and rewarding to read as I have found it to research and write.

11

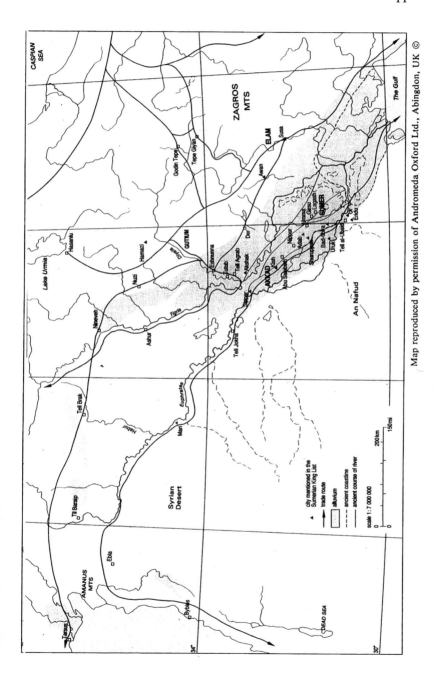

Map reproduced by permission of Andromeda Oxford Ltd., Abingdon, UK ©

Chapter 1

ADAM'S HISTORICAL NICHE: WHO'S ON FIRST?

"Well, I guess this shoots down Adam and Eve," the young man turned to his companion and mused.

"I know," she replied, "but it's such a beautiful story."

I overheard these remarks in the summer of 1984, at an exhibition held at the New York Museum of Natural History entitled, "Ancestors - Four Million Years of Humanity." A special collection of the fossil remains of early man was on display gleaned from private and state-owned museums from all over the world.

What prompted those comments by the young couple rested in a glass container in the center display area. There sat the daunting fossilized skull of an *Australopithicus africanus* estimated to be 2 to 3 million years old. With his prominent, gorilla-like brow ridge and small brain case, he stared out at his audience with unseeing eyes.

I yearned to interject.

That is more than a story, I wanted to say! It is a foundation stone of Scripture. Adam and Eve's creation was recorded by Moses, a godly servant who did not blunder in what he compiled, inasmuch as he was protected from error by a God who does not lie. Ah, but those old bones, were they the liars then?

The Battleground

The origins question has been a sticking point for generations. Did we arrive here by way of creation or evolution, and how can we be sure? One key reason the issue of creation versus evolution is so controversial, dare I say explosive, is that both sides have taken uncompromising positions.

We are not talking about a few minor interpretative differences between proponents of two credible theories, but two armed camps, poles apart in thinking, with no hope or even desire to meet in the middle.

Antagonists who stake their reputations on either sudden creation or gradual evolution have pitted night against day with no room for concessions, no prize for second best. These extremist proponents do not seek common ground where their positions could be reconciled.

Dogmatists on both sides of the argument have held fast and wielded brickbats at their opponents unmoved by a wealth of contrary evidence which, if only they took time to review it, shows that each side has built its case partly on credible evidence and partly on shifting sand.

Whereas most evolutionists disavow any possibility of intelligent design in the evolutionary process, young-earth creationists ignore evidence that natural processes have operated in our universe for billions of years.

And while young-earth creationists try to force even the Bible to succumb to their ill-conceived conclusions, many evolutionists disdain God's existence, and assign all of biological history to a materialistic theory governed by "order in the universe" and "random chance."

On the positive side (if there is a positive side) creationists recognize the undeniable Creator and see His indelible stamp. God is not vacationing out in the universe, preoccupied in a distant galaxy. He is a personal, loving Father who has numbered the hairs on our head. Life on earth and life hereafter are attributable to His love and His grace. There are no accidental or unforeseen processes in God's creation.

Evolutionists, at the other extreme, are bound to the practical application of scientific observation. They compile and interpret data. The work of paleontologists is confirmed by archaeologists and taxonomists. Conclusions reached by biologists are supported by geneticists and geologists. Physicists concur with astronomers and chemists. Scientists from all areas are virtually unanimous in subscribing to the descent of complex animal life including man from simple life forms, the main tenet of the theory of evolution.

Are eminent scientists part of a deceptive scheme, or are they bound by a strict code of ethics? Must they adhere to well-prescribed methods and procedures in recording data and reporting their conclusions, or are they allowed to run amuck, operating with carelessness and reckless abandon?

By and large, you would discover that those in the scientific community, whether practicing science in research, in industry, or in a teaching role, investigate thoroughly and corroborate exhaustively before their findings appear in print. Rival colleagues and competing companies and laboratories are ever anxious to pinpoint any discrepancies or deviations from accepted practice in competitors' theories, procedures, or conclusions. Professional reputations - and along with it, compensation, promotions and grant money - can hinge on the accuracy and thoroughness of the work.

It is not the sincerity or dedication of anyone who has taken a stand on this most difficult of issues, whether creationist or evolutionist, that is in question here. Judgment is another matter. It is all too obvious that fallacy, and even falsity, has crept into the rival camps. And just as Pinocchio learned to his chagrin, it has become "as plain as the nose on your face."

Choosing the Gate

"The audiences that come to hear me are, I often feel, seeking a kind of reassurance," states the distinguished paleontologist Richard Leakey in the prologue to his book *Origins Reconsidered*, referring to the question and answer sessions that close his public lectures. [1] And he is right. They do need "reassurance." If the philosophical undercurrents of evolutionism are legitimate, this may be the escape clause freeing us from any demands of God, and, of course, any penalties.

If we are really nothing more than clever, bipedal apes, a major premise of Leakey's books and lectures, then we can kick back and relax. As the temporary end product of a long string of fortunate accidents, this could be cause to rejoice. We have a "Get Out of Jail Free" card. We can sing "I Did It My Way" Sinatra-style without irritating repercussions. Shunning the "narrow gate" (Matt. 13,14), we can sashay through that "wide gate" doing a soft shoe, wearing a top hat, spats, and cane *if* the evolutionists are right.

What if the evolutionists are wrong? The stakes are quite high. The Bible is replete with ominous warnings. If we choose the wrong path, we face "weeping," "wailing," and "gnashing of teeth" (Matt. 8:12; 13:42; 13:50; 22:13; 24:51; 25:30; Luke 13:28). And it only lasts for all eternity!

On the positive side, life is in the offing - eternal life, and although there are some conditions, it is available to all. How tragic for this gift to lie unclaimed because of a simple misunderstanding by sincere individuals dedicated to honest science on one hand, and those committed to God's infallible Word on the other.

Searching for Common Ground

These are the goals: (1) to acknowledge and strengthen the truth wherever it is found; (2) to expose the errors and falsehoods that are interwoven into both of these competing explanations; and (3) to show conclusively that common ground can be established, there is a solution.

Theories, or hypotheses, have been offered by the Christian community that take a middle position between the extremes of atheistic evolutionism and young-earth creationism. These are Theistic Evolution (okay, evolution - but God did it), Gap Theory (destruction followed by re-creation), and Progressive Creation (intermittent acts of special creation with micro-evolution in between). These proposed solutions originate from well-intentioned attempts at finding common ground. We will examine these alternate explanations and explore their inherent weaknesses.

Once the decks are cleared of proposed solutions, both the unthinkable and the unworkable, the answer detailed in this book is shown to solve the stickiest problems associated with how we got here, if we can depend not only upon the truth of the Bible, but also upon the validity of current findings in history and science. Notwithstanding, the subject of human origins is an issue involving more than just gathering and interpreting data.

The origin of life and man has been a divisive issue that arouses passions and foments intense competition among rivals advocating different schools of thought. Egos, jobs, endowments, prestige, and tenure can be obtained, enhanced, jeopardized, or lost on the basis of whether one advocates creation or evolution.

At the same time, some will say what difference does it make? Who cares how we got here anyway? It is what we do now that we are here that is of paramount importance, is it not? Why are we so preoccupied with who our ancestors might have been?

Laying Traps

I had spent some time discussing the origins question with a colleague of mine. We share common beliefs of God's role and man's place in the scheme of things. One day as I walked in the door of our shared office, he looked up from his typewriter and started with, "You know, it really is a moot point." (Why waste time with salutary niceties.)

You know, you're right," I responded. "It's our faith and obedience which is truly important. But how do you reckon with the Genesis account of Adam and Eve?"

"Oh, that's simple," he grinned. "I just don't believe it."

That's a tack many well-intentioned Christians have taken. They ignore the issue completely and subscribe the entire problem to either Bible allegory, Bible poetry, or Bible error. Apparently, Moses was daydreaming, or he was reporting a Babylonian myth which was popular in those days. The important stuff is found in the New Testament anyway, so why worry about it?

Is this the right attitude toward Scripture? Should we not hold both the Old Testament and the New Testament in high esteem? After all, the New Testament builds on the foundation of the Old Testament. If the old covenants are rooted in mythology, how could we comprehend fully or place our full confidence in the new covenant? If the "first Adam" (1 Cor. 15:45) never lived, what would be the implications for the "last Adam" - the one we call Christ?

The Pastor of a Bible-teaching church in Virginia, just outside Washington, D.C., is a staunch young-earth creationist. He ascribes to special or "fiat" creation, which presumably took just six 24-hour days from the birth of the universe to the creation of modern man. And yet he is only one among thousands of evangelical Christians who have taken this ultra-conservative approach. They take no quarter with conciliatory attempts at compromise with a sinful world's godless explanations.

Stepping on the toes of modern science in the name of God is only half the sorry equation. What is worse, these advocates of young-earth creationism present tortured interpretations of the very Scripture they are defending in order to support their claim. Must bewildered believers put on blinders to accompany their faith? Can there not be a reasoned approach to a literal interpretation?

My father was reared in the faith. He and his eleven brothers and sisters attended a small-town Catholic church in Nebraska. While enrolled at the University of Omaha, he took a required biology course where his professor revealed that God took no part in bringing man about.

He was told without equivocation that man evolved from monkeys. Until his death, my father rejected the Bible, clinging to the wisdom of man over the wisdom of God. How many others have been cheated and shortchanged by the philosophy of evolutionism as proclaimed in secular and even in some Christian academic institutions?

A man who is an author and professor of biology has written an authoritative critique of what is called "creation-science." In his book he exposes the weaknesses in young-earth creation theory with the skill of a surgeon cutting out a cancerous growth. He dispels creationist thinking and defends evolutionary theory with sound arguments based on a preponderance of scientific evidence.

In his dialogue he includes a philosophy of unforeseen chance and purposeless life. What is the purpose of life, he asks, or to what purpose is a "tapeworm"? All of life, according to this biology professor, is attributable to "impersonal natural laws." 2 Is it necessary for educators to try to shoehorn atheism into our brains along with teaching science?

An author of a popular book on physics ventures that it is more reasonable to suppose that our marvelous universe is self-caused, self-originating, and self-organizing than it is to believe that a supreme intelligence might have been responsible. [3] Are we to believe a burning bush *itself* spoke to Moses (Exod. 3:2-4)? What would be the physics of that?

These are just some of the clever snares set to trap us as we try to negotiate our way through an avalanche of information and misinformation. To theists, the Creator must be responsible, and an integral part of any theory which brought man into the world. To scientists who subscribe to evolutionary processes, any theory necessitating divine intervention will be discarded out of hand.

I empathize with those who have been caught up innocently in one side or the other of this difficult dilemma. If this book does nothing more than to bring into vogue a one-phrase response - "I don't know" - for pastors, laymen, educators, and bystanders when queried on the subject of man's origins, then it will have done its work. There is no dishonor in sidestepping the issue. As the late evangelist Dr. Donald Barnhouse would say, "Your problem is not evolution. Your problem is sin."

For those who demand answers, though, they will not come away dissatisfied. The solution to the creation-evolution dilemma put forth herein is consistent with historical, natural, and biblical evidence. It is a workable, testable answer to the question of man's origins if the word of God is true in the early chapters of Genesis in a literal sense, and if the most recent findings of science are to be taken seriously.

House of Science

Whether the answer advocated in the following pages will endure the ages and stand up to ultimate scrutiny - well, who knows? But for the present, let us liken it to a key to the front door of your house.

Have you ever fumbled with a set of keys while standing outside in the cold at night trying to figure out which is the right one? Each is tried without success until eventually the right key slides into the lock and - Aha! A few seconds later, you are enjoying the warmth and coziness of your home. Next time leave the light on.

Now let us imagine the house in which you live is the world of science. Your home is comprised of the data-filled environment surrounding you today. Each room depicts a separate discipline. The dining room is the science of biology. Geology is the living room. For bedrooms, you have physics, astronomy, archaeology, and so on. Each room impacts your life as you enter and examine its contents, but without heat it would be cold and dreary.

A benevolent God furnishes the heat through His love and blessings. Your home can be enjoyed only in God's warmth. Without Him, no real peace can be found. Oh to be sure, you can enjoy your house at times of the year when the weather is warm. But only heat offsets the dread cold of winter which is found inside the door that has a lock and needs a key.

What if none of the keys fits? The illogical, unworkable, and entrenched answers currently in vogue are merely ill-fitting keys that fail to unlock the door. Young-earth creationists wrench a twisted nail into the lock's opening and find the door unyielding. So they huddle in their car with the motor running. At least they stay warm. But by doing this they spurn a great body of knowledge about the world in which we live.

God-denying evolutionists kick the door down, let the heat escape, and try to live in a freezing house. In life, there are times we can go it alone. Our society encourages and rewards self-reliance, but this leaves us nowhere to turn in times of need. So the means to proper entry are available in our house of science, but at too great a cost. We need a workable key. The answer offered herein is simply a key that works on a ring of nonfunctioning alternatives.

Setting the Stage

A modest amount of faith can be helpful, but faith alone is not the answer. Cults and false religions, based entirely on faith, wither under scientific scrutiny. True Christianity stands out as the only religion equipped with evidence to back it up. If all of us would enter into an exhausting search of the available data, while setting aside old prejudices, here are six conclusions probably most of us would reach:

1. God lives! Everything that is, was, or ever will be, owes its existence to His creative power.

2. The Bible tells us of His nature, His relationship to us, how He wants us to live, and what we must do to attain His kingdom.

3. The Books of the Bible are trustworthy, authoritative, and in the original versions, inerrant. Furthermore, except in obvious instances where a story or parable is being told, or a metaphor is used, the Bible should be taken literally.

4. Our Creator has put in place certain natural causative factors in sustaining life and bringing about adaptive changes in life forms.

5. Scientists, who study nature, earned their degrees through years of hard work, diligent study, and careful research. They have contributed enormously to man's health and well-being, and our growing body of knowledge.

6. General revelation provided by observing nature, and special revelation that comes from Scripture, should be complementary since both are from the same author - God. If Scripture and nature appear to be contradictory, it is due to our propensity to misinterpret and misunderstand.

What to Expect

The religion that is married to science today will be a widow tomorrow. The sciences in their multiple theories and forms come and go. Biology in the year 2050 may be as different from the biology of today as the religion of today is from the religion of 1850. But the religion that is divorced from science today will leave no offspring tomorrow. 4

What should we expect of a theory which purports to resolve the heretofore unresolvable? A Bible-based explanation should give the Creator full credit for causing all forms of life, including man, to come about. The creation narrative, as set out in the first chapters of Genesis to include Adam and Eve, temptation and sin, must be given full weight if such a theory is to be accepted by the evangelical community.

At the same time, scientists will never accept any theory of reconciliation unless evidence is taken into account that bipedal creatures of some description have been walking upright on this planet for something like four million years or more. Thanks to relentless pursuit, scientists have established the antiquity of man and his precursors. An acceptable solution must recognize ancient human ancestry. Also, any such theory should not crumble if a genetic link is ever proven conclusively between man and other higher primates.

The solution tendered in the following pages recognizes the existence of prehistoric man. Those old fossils that have been unearthed are not part of some demonic plot to lead us astray. Early man not only existed, but was well established long before God introduced the first covenant human being, a man called Adam.

The majority of Bible-sensitive theories relating to the origins of man lack internal consistency. This can also be said of many Genesis commentaries. It only adds to confusion to take a stance on one particular Scripture verse or scientific observation, and take a contrary position somewhere else. The solution advocated herein honors God's Word, gives full weight to scientific achievement, and it is internally consistent.

A New Approach

It will be demonstrated from the words of Scripture and confirmed by the testimony from nature that Adam did not start our species, but was inserted into an already populated world. We can establish not only the place and time of Adam's appearance with a fair degree of certainty, but also there are clues concerning the surrounding culture.

According to Scripture, Adam was the first man with whom the Creator established a special relationship. This unique status does not mean he was the first biped, the first hominid, or first of the genus *Homo*, or even first of the species we call *Homo sapiens*. Adam was the first, however, to have a contract which offered eternal life conditional upon his obedience.

God's only requirement of Adam to maintain close communion was to avoid eating fruit from a certain tree. It does not sound difficult, but it proved to be the stumbling block which led to his and eventually to our downfall.

Adam was not conceived by earthly parents according to the Bible. The Lord God formed him out of the ground. Eve was fashioned from Adam's side giving her a unique origin as well. Many scoff at this notion and try to relegate the Genesis account to poetry, mythology, or allegory. It will be demonstrated, however, that the narrative pertaining to Adam and Eve need not be modified, discounted, or explained away due to scientific discovery.

Just as Adam was placed into a populated world to bring the news of God's kingdom to the polytheistic, earthly inhabitants, likewise Jesus entered at a later date for the same purpose. God chose a special human vessel, Mary, from whom was born the Messiah to the Jews, later to become the Savior to a sin-plagued world. Christ was singularly unique, yet like Adam and Eve, there was no human father.

Creation Methods

Through the descriptions of the origins of Adam, Eve, and Christ, the Bible describes three ways in which the Creator may implant directly into the human race. Conceivably, God could have used any of these methods to introduce new creatures into the world's life stream, had He chosen to do so. In this context the word "implant" simply means that God has the power to bring new life into being in an extraordinary manner if it suits His purpose without two biological parents.

The methods of implantation God used as set out in the Bible are as follows:

1. God may have formed man or animals directly from the "dust of the ground" without genetic links to previously existing species (Gen. 2:7,19).

2. He can fashion one from part of another as He did with Eve (Gen. 2:21-22).

3. A female may bear an offspring without the normal fertilization process (Matt. 1:23, Luke 1:35); however, the incarnate Jesus may have been the only case where this happened.

A fourth possibility exists, though not implied in Scripture. Natural parents, through the normal mating process, may have had natural offspring, but changed through the mechanics of DNA modification. That too would be a creative act, if God worked that way, and some think He may have.

We then have three or four methods God could have used to bring into existence human beings, or animals, as it suited His purpose. Of course, God may have used or allowed what we commonly call evolutionary processes.

One point should be emphasized. There are no limits to what God can do as far as we know, except that there appear to be limits He has placed upon Himself. He cannot lie, for example (Titus 1:2). He may have used some manner of evolution to the despair of creationists, and He may have used various means of creation to the dismay of evolutionists.

Although we are free to speculate on where or when God used creation or evolution, we have something available to us so that we can know with a fair degree of confidence. It is called data. And it comes from the Bible, and from science and history. Facts take precedence over theories, even theories conceived brilliantly. No matter how well-intended the theory may be, if the facts are contradictory, it fails and should be discarded. It is as simple as that.

A Man and a Woman

In Genesis Chapter 1, a man and woman were created in the "image of God," in His "likeness." Some Bible scholars have extended this verse to encompass all of mankind. The traditional interpretation, and probably the correct one, is that this verse refers specifically to Adam and Eve, although Adam is not named until the next chapter.

Genesis 1-2:3 differs markedly from the rest of Genesis, and it may be that two narratives, from different sources, were put together or compiled by Moses. It has been advocated that the creation of man in Genesis 1 should be separated from the subsequent creation account beginning at Genesis 2:4 to put a length of time between the creation of mankind and the creation of Adam.

This "gap theory" was advanced by Thomas Chalmers in the early 1800's. [5] C. I. Scofield breathed life into this methodology in his Reference Bible published in 1909. The precursor to archaic man was named "Adam 1" from Genesis 1:27 to start the human (or subhuman) race. This would account for the previously unaccountable fossilized remains of early man. According to Scofield, modern human beings, including everyone living today, emanated from the Adam described in the second chapter of Genesis.

Scofield's method sought to avoid the difficulties caused by the appearance of Adam of Genesis 2 at a point in history too late for him to be ancestral to every fossilized bone that has been dug up and assigned to the genus *Homo*.

In this gap scenario, early precursors were eradicated by some natural calamity, or died out because of ill-suited survival techniques. They were replaced miraculously and divinely by a whole new species more modern in appearance. Many Bible scholars have pronounced this method exegetically unsound, but Scofield did take a needed step toward reconciliation.

A man and a woman, or men and women, brought into existence through either a creative act or an evolutionary process, have a definite niche in history somewhere in the neighborhood of 200,000 to 100,000 years ago. That early couple, or those couples, were the first modern *Homo sapiens*; ultimately from them came the great races covering the globe today. If this was an act of creation, it could have been accomplished simply through DNA modification of selected members or a selected member of a particular population of ancient hominids. Or perhaps, subtle, gradual DNA modification through long periods of time may have furnished a smooth transition from one species to the next.

Genetic Links

As to the question of whether present-day human beings are biologically or genetically linked to apes, the answer should not cause dismay for even the most conservative-minded evangelicals. Man is both unique and yet remarkably similar to the gorilla and chimpanzee. To presume the absence of any link or bridge between human beings and other primate species runs counter to what virtually every biologist recognizes.

Although the idea may offend some innate sensibilities, the solution tendered herein demonstrates conclusively that ancient non-human forbears cannot be denied by way of inference from the biblical narrative. We can deny genetic linkage on the basis of philosophical prejudice if we like, but the truth of the matter we will see, is that the Genesis account neither confirms nor denies non-human ancestors.

At this point in human knowledge, scientists cannot prove absolutely that a genetic link exists between man and other members of the suborder Anthropoidea which includes chimps and gorillas. If God chose a point 4 to 5 million years ago to create specially or insert a bipedal precursor to man into a world already populated by lesser animals including apes, the only question would be, what method did He use?

If God used creation method number one discussed previously, using the lesser apes as a kind of model, then He might have brought early man into the world by forming him out of the ground. In that event, man would be unique, and unrelated to any other species. He would just look sort of apelike.

Removing a rib from an ape, and fashioning a man from it is a bit gruesome, so let us skip to the third method.

Was a particular female ape singled out to produce a miraculous offspring - man himself - without the participation of any father ape? Who knows the answer to that question? And the apes are not talking. So on to method four.

Perhaps, by the natural fertilization process, two sets of ape parents had two strange looking offspring (strange by ape standards) who took a liking to one another and eloped to a faraway place. Could they have produced progeny who became human eventually, through DNA modification? Yes, it is possible, and furthermore, if it happened that way, it would have exactly zero impact on the historical accuracy of the creation account in Genesis! That precisely is the point of this book.

A Long Procession of Hominids

Regardless of the creative method or evolutionary processes employed, from the first bipeds, or human forerunners, came beings whose bones keep popping up in various locations. *Australopithicus africanus* could have been a distant relative. *A. boise* and *A. afarensis* may have been offshoots. *Homo habilis* and later *Homo erectus* could have been their descendants leading to archaic *Homo sapiens*. It is assumed that modern *Homo sapiens* descended from archaic types.

The Neanderthals, who began some 130,000 years ago and died out approximately 35,000 years ago, may have originated before modern *Homo sapiens* began to spread out about 110,000 to 50,000 years ago. Or Neanderthal may have been a descendant, an aberrant offshoot, that became extinct. It may be that the smaller *Homo sapiens*, in search of new worlds to conquer, helped push Neanderthal and others into extinction. Further research may clarify that in time. In any event, none of us appears to be descended directly from Neanderthals.

Cro-Magnon man, whose remains have been found primarily in caves in France, dates to 35,000 years ago, and appears to be fully modern. Bear in mind that archaeological evidence is still sketchy. Dating methods are revised as new techniques are discovered and old ones improved. The point is this: modern *Homo sapiens* had precursors - that is a fact; and they may have been our ancestors - that is a theory.

Bending the Text

To be sure, liberal interpretations of the Old Testament taught at many centers of Bible learning today will not have difficulty with any unearthing of the remains of early hominids or with any other future scientific discoveries. But the current trend toward a liberal interpretation of Genesis and the rest of the Old Testament sprang up partly due to the difficulties which have persisted between science and Scripture in the first place.

A liberal, relaxed view toward the Old Testament is unnecessary. We need not discount the early passages of Genesis to make the Bible palatable in a science-minded world. What is needed and presented herein is an enlightened, *literal* interpretation of the first eleven chapters of Genesis, not an appeasing interpretation dictated by scientific enlightenment.

A conciliatory interpretation of Genesis, subservient to human rationale, may incline its followers to subrogate the entire Bible to the whims of human understanding. By denying historical accuracy, well-meaning apologists have unleashed a dragon bent on devouring the entire Old Testament - a beast that munches on Matthew, and nibbles on the rest of the New Testament. If the opening passages of Genesis are not historically accurate, at what chapter, or starting with which Book of the Bible, should we regard the Scriptures differently?

Reconciliation

A giant step toward reconciliation between Scripture and science is long overdue, at least as it pertains to the origins of life and man. The purpose of this book is to remove obstacles, expose myths, embarrass demagogues, reveal truth, promote further serious inquiry, and finally solve a problem that has been a pitfall of immense proportions for scientists and theologians for centuries.

This book has been written for the Christian parent whose son or daughter is enrolled in Biology I; for the geology teacher when asked how the text book could be accurate when the earth is only

6,000 years old; for the pastor who is called to account for the obvious "errors" in Genesis; and for the student or layperson who only wants to know how to make sense of it all. In short, this is for those who are seeking an honest answer amidst a clamor of conflicting opinions, as well as for those who were not looking, but are glad to have an answer in case they are ever asked.

Years ago, I saw a small segment of a movie that was playing on TV. A Middle Eastern philosopher or teacher was instructing a young lad to develop his powers of observation. Briefly, the old man showed his student a tray holding many precious and semi-precious stones while telling him it held five Cats Eyes for example.

After the tray was removed from view, the boy was asked to tell the number of each variety of stones he had seen. Even though he had seen the tray for only an instant, he told his mentor the exact number of rubies, sapphires, emeralds, and the like, ending his recitation with, "and five Cats Eyes."

"No, you're wrong," corrected the old man. "There were six Cats Eyes." The boy was astonished. Then the teacher revealed the real lesson he wanted the youngster to learn, "Never trust to the word of another what you are able to see for yourself."

In the 1950's an upstart electronics firm had a slogan that was adopted by many who were not employed by or associated with that company. Just one word was inscribed on plaques that took center stage on many executives' desks. The company - IBM, the word - "Think."

Have faith and believe? Yes, but we can't stop thinking, and we can't rely on others to do it for us. Thinking is part of what this book is all about.

NOTES

1 Richard Leakey and Roger Lewin, *Origins Reconsidered* (New York: Doubleday, 1992), xvi.

2 Douglas J. Futuyama, *Science On Trial: The Case for Evolution* (New York: Pantheon Books, 1983), 37.

3 Paul Davies, *God and the New Physics* (New York: Simon & Schuster, 1983), 49.

4 Holmes Rolston, III, *Science and Religion: A Critical Survey* (New York: Random House: 1987), vii.

5 Thomas Chalmers, "Remarks on Cuvier's Theory of the Earth; in Extracts from a Review of that theory which was Contributed to the Christian Instructor in 1814," *The Select Works of Thomas Chalmers* (New York: Robert Carter and Brothers, 1850), vol. 1, pp. 180-193.

Chapter 2

THEY CALL IT SCIENCE:
A ROSE BY MANY NAMES

In the early stages of World War II, a crippled British Lancaster was returning from a midnight bombing raid on Nazi Germany. The valiant crew nursed their war plane back across the North Sea in a desperate effort to reach friendly soil. Lower and lower they sank until they were barely skimming the waves. Finally, just minutes before they would have made landfall, the crew was forced to ditch, belly landing in the black water just a few miles from the Scottish coast.

None of the crew was injured in the crash landing, all escaped safely, but there was not enough time to launch a life raft from the rapidly sinking plane. An overcast sky accompanied by a total blackout, part of the defensive wartime posture, gave no clues to lead them to safety. So they floated together, supported by their leather bomber jackets, and waited patiently for the first rays of sunlight to guide them to shore. Then, what had been life support became life threatening.

Although flotation cartridges were sewn into each leather jacket, what must surely seem odd to us now, they had not been tested! The jackets floated at first, but after hours of submersion, the leather soaked up water and became leaden. To their horror, the buoyant jackets turned to deadly anchors. Before daybreak every crew member had been dragged beneath the waves except one.

The navigator, John Haffenden, had been a champion swimmer as a boy. Shedding his water-logged flight gear, he tread water slowly until the first rays of sunlight coaxed the faint coastline out of the mist. Then he began to swim. Stroke after stroke he labored to reach shore seven miles from the ditched plane. At one point a cramp seized his calf. Treading water with one hand, he massaged his leg with the other until the cramp began to ease.

On he plodded through the water until cramps struck both legs. The pain was more than he could bear this time. Still short of the coast, he gave up and prepared to drown. Closing his eyes and lying back, slowly, he let his feet sink. They touched sand!

Though still short of land, he stood on a sand bar with his head barely out of water until the cramps eased. After a brief rest, he resumed swimming until eventually he reached shore. Weary and trembling with cold, he clambered out of the sea only to encounter another peril. The beach was mined.

As if land mines were not enough of a hazard, this particular stretch of beach was a bird sanctuary, and this happened to be nesting time. Not recognizing him as anything special, the alarmed birds treated this RAF navigator like any other predator and began dive-bombing him. With no time to sit around and get pecked to pieces, he scrambled zigzag fashion up the beach, scooping up handfuls of sand and tossing them over his back to shoo away the attacking birds dreading an explosion with every step.

After clearing the hazards of the beach, Haffenden spotted a coastal lookout post in the distance, and off he hiked across the rocks and sand. An elderly coast guard volunteer had just begun his watch, and was brewing a pot of tea to warm himself and bolster his morale for another day of dreary solitude. Suddenly the door opened, a man wearing only a wrist watch stood silhouetted against the gray morning sky, and voiced a quiet request, "I'll have a cup of that, if you don't mind."

Haffy, as he was called affectionately, received a hero's welcome. RAF bomber crews were issued American "Mae West" flotation gear they used for the balance of the war. When a navigator was needed to transport Winston Churchill to Yalta to meet with President Roosevelt, John Haffenden was chosen.

After the war, Haffenden married and started a business in London. His son, Raymond, grew up there and became part of the company. He married and had two sons.

Now, let us ponder for a moment. To what extent do Raymond and his sons owe their very existence to a sand bar placed conveniently in the North Sea? That incident in World War II is but one event in a gigantic spider web of interconnected events that impacted their lives, and mine, and countless other lives as well.

Dice or Deity?

Who among the living today have fathers or grandfathers who were RAF bomber crew members, and floated in the ocean waters supported by Mae Wests, and thanks to American ingenuity, survived the war? Who would have been born, had it not been for the untimely death of their potential fathers, and thus were deprived of the opportunity for life?

Was Haffy simply fortunate in contrast to his unlucky crew members? If luck was involved, then are the offspring the fortuitous results of a roll-of-the-dice process? How many other, seemingly unrelated, fortunate events have taken place to bring into being favored creatures who only got here because of accidental events throughout eons of time? If that is the answer, then let us set up an idol, call it Serendipity, and offer our gifts.

Perhaps destiny guided John Haffenden to a submerged sandy plateau that shifted into place just so by capricious sea currents. Or was the sand bar itself shaped by invisible forces planned in advance for a rendezvous that would save his life and allow for future lives? Was he destined to live? Were the others fated to die? If so, why? What has been predetermined, and what has been left to chance?

Was there a particular reason why this one man escaped death? Was it for the benefit of his unborn son or one of his subsequent descendants who will some day father or mother a future prime minister? Was some special person the express reason for a series of unseen interventions making this event just one of many?

Were the winds of fate blowing in the North Sea on that day for the benefit of someone destined to play an important role in the future - a role which would have been thwarted had the Nazi gunners achieved total success? Was there a purpose known only to Providence that required intercession?

Did God, Himself, go out of His way and take special effort to guide one particular RAF crew member in an act of mercy? Was the fate of the entire crew directly in the hands of a sovereign God, who decided on the spot who would live and who would die? Or had the die been cast since the beginning of time? Was it that Haffenden lived and the others died because otherwise, there would have been no sons or daughters, who had to be born, since God knew them since before time began?

Did God's hands perform a specific act of clemency and perform a miracle to ensure future events? If this is the answer, then God must be an interventionist who interacts from time to time so that what He wants to happen will happen. In between would be just free-wheeling.

An Omniscient God

Instead of the whims of chance, fate, or sporadic divine intervention, let us look at one of God's characteristics for the answer. We know that God is omniscient or all-knowing. Matter, space, time, and energy were created by God, but He is not governed by, nor bound by, His creation. Time is within God's domain, not the other way around. Thus, God knows the end from the beginning and all the events in between.

We are bound by time. Completely subject to past events, we are locked in the present, and experience the future moment by moment as we move through space-time. God is not so encumbered. He transcends time, the future is as clearly visible and knowable to Him as the present is to us.

> In the mind of God, there is no "before" or "after"; there is only a "now." In "God's experience" all events occur simultaneously. To put it another way, all the past and all the future "that is, *our* past and future" exist together in God's present. [1]

God does not push us along with constant or intermittent intervention to get us where He wants us to end up. Seeing the complete rainbow of time, Haffy's offspring were as known to God when Haffy stood on the sand bar, as they are today.

This does not negate miracles. God is quite capable of performing miracles, but even miracles are foreknown. In essence, we live in a miracle. This universe and the world we live in is a wonder of God's creation. He initiated a miracle, and sustains the miracle, according to His divine Will. Life itself is miraculous, but it is not necessary to call for divine intervention at every juncture. It may seem paradoxical at first that we can believe in God's wonders, but need not invoke miracles in the procession of life. Yet this is no different than the seeming paradox of predestination versus free will.

As the First Cause of all events, this is not to say that God is the *causer* of all events. His knowing the future does not preempt our free will. We are given the freedom, without divine control, to make our own uncoerced choices from among any number of alternative courses of action. We choose from available options, and make our decisions, either in His Will or against His Will.

We alone decide, guided by intellect, circumstance, habit, instinct, advice, prayer, etc. If we make bad choices we endure consequences, although God can make good come from bad. So we may suffer. He does not.

When we select proper choices God desires for us, we prosper and receive blessings. Does He agonize over the choices we make, wondering what we will do from moment to moment? No, the choices we make God knew about before the beginning of time.

That British war plane could have crashed hundreds of miles from shore. Haffenden could have commenced swimming in any direction; his legs could have cramped anytime; he could have given up at any point; the sand bar could have been anyplace. Yet, it was all foreknown. There was no necessity for God to take specific, miraculous, or intermediate actions along the way. The events simply moved along a time line that was already known to an omniscient God.

In the words of Paul, "According as He hath chosen us in Him before the foundation of the world ..." (Eph. 1:4). Also from Romans 8:29: "For whom He did foreknow, He also did predestinate to be conformed to the image of His Son ..." Considering all of us were known to God from the beginning, then the events that caused us all to be born into the world also must have been known.

An example of God's omniscience can be found in I Kings 22. Jehoshaphat was requested to join the king of Israel, Ahab, in battle against the Syrians. Prophets were queried so that Ahab and Jehoshaphat would know the results beforehand. Micaiah was known as one "by whom we may enquire of the Lord" (I Kings 22:8). It did not bode well for the king of Israel that the prophet Micaiah saw all of Israel scattered about as sheep without a shepherd. Undaunted, Ahab thought of a clever scheme to thwart the prophecy.

He persuaded Jehoshaphat to dress as if he were king. Jehoshaphat obliged and they went into battle. The Syrians sought to slay Jehoshaphat thinking he was the king, but when they discovered the deception, they turned back. Meanwhile, Ahab went into battle in disguise, and wouldn't you know it, he was slain when an arrow shot at random struck him at the "joints of the harness" (I Kings 22:34).

Some might say that although the arrow was shot randomly, God seized the opportunity, intercepted it in flight, and guided the arrow to its mark in order to fulfill His divine Will. Did God guide the arrow? Was an act of intervention required to cause an event to take place which otherwise would not have happened had God not acted sovereignly?

Those who say "yes" subscribe to interventionism. According to this idea, God intervenes frequently or continually in the course of history to bring about desired results. If God is omniscient, however, and knows the future already, why would He have to engage in continual corrective action?

The prophet Micaiah, acting on request, gave the king a part of God's foreknowledge. God had already foreseen the prophet being asked to tell the outcome of the battle. God foreknew the king would die. He further knew that the king would die in disguise with an arrow piercing the joints of his harness, though the prophet did not go into additional detail. The end of the king was seen from the beginning. The inevitable outcome had been pre-known by an omniscient God - intervention not required.

As it pertains to evolution, it is this notion that God must intervene continually to bring life into being that sets creationists at odds with most scientists. Calling for a divine miracle wherever a gap occurs in the fossil record, for example, runs counter to what scientists see as a purely natural process. And as we have perceived that God does not need to intervene, we can set interventionism aside as unnecessary to explain the procession of life.

Science as a Process

Let us look at science. First, we will deal with semantics. One of the key problems clouding the origins issue is that words like "science" and "scientific," "evolution" and "evolutionism," "creation" and "creationism" are thrown around as if everybody already knew precisely what the words mean, and therefore, do not require defining. This is a principal source of misunderstanding perpetuated by many who call themselves creationists, as well as those who embrace evolution.

In addition to science consisting of a body of knowledge, it is also a process. Science is a method or system by which we make observations, gather data, formulate theories, arrive at conclusions, etc. A method can be labeled "scientific" if it adheres to certain commonly accepted rules that have been established to insure accuracy.

We might make an observation, or witness a phenomenon. We may be able to make repeated observations and record precise details as to what we observed, although often times we do not have that luxury. We can make wild or educated guesses as to the cause of a particular phenomenon. Plain old guesswork may be used in the initial phase, and trial balloons sent up.

The next phase may involve gathering data, or conducting experiments. Using observations or experimental results, some of the derived data points may be connected up yielding one or more working hypotheses, which can then be tested. Some initial suggested answers usually are found wanting, and are discarded. Normally one or more of the working hypotheses garners support as other researchers begin to accept it as plausible, elevating it to the status of theory.

Although a working hypothesis connects up some, most, or all of the data points, it needs to attain a measure of acceptability before it can be called a theory. Theories can be tested by themselves, or against one another. When one theory gains virtually universal acceptance it becomes a law.

In 1982, the solution detailed in this book was mere conjecture. By 1986, when *The Washington Post* published an article outlining this proposal, it could have been called a working hypothesis, even though the article proclaimed it a theory prematurely. The method of reconciliation between science and the Bible presented in this book can be called a theory today considering the supporting evidence.

If general acceptance is gained, displacing all competing Bible-sensitive explanations in prominence, this theory may gain status as a paradigm in the Bible-believing community. Among scientists and academics, biological evolution is the commonly accepted paradigm today, that even withstands contrary evidence as long as no superior solution can replace it.

At the opposite end of the spectrum, opposed to theories, paradigms, facts, and laws, we encounter fallacies, falsehoods, and lies. Advocated solutions put forth in spite of the data, contrary to the evidence, may be assigned to one of these three categories. We can apply the distinctions according to how grossly they violate accepted scientific practices. Fallacies defy conflicting data, falsehoods deny the facts, while normally, lies have evil intent.

An acceptable explanation should be falsifiable. That means there must be a way available to try to disprove it, usually through some testing procedure or series of observations. If there is no possible way to invalidate something, it falls into another category, such as an undeniable fact, an illogical conjecture, an outright lie, or a bona fide miracle.

By their very nature, miracles cannot be proven. Thus miracles fall outside the purview of the scientific method. In God's interaction with humans, miracles go with the territory; that is part of what being God is all about. He has set up a boundary condition for us, however, which He honors as well.

In all instances in Scripture when God stepped outside the boundaries of scientific explanation, He did it primarily to demonstrate His capacity to so act, or prove the authority of those acting on His behalf: Christ, the prophets, the apostles, and the body of believers. Outside of miracles performed for such a purpose, we can depend upon laws of physics to be in operation and totally predictable all the time.

God did not give us an "Alice in Wonderland" world that would keep us in continual disarray. We can live without consternation in a predictable habitat, and not have to worry on a minute by minute basis what some unpredictable, capricious deity has in store for us. Miracles are still possible. In fact, miracles are part of our belief structure, if we trust the Bible, and put our faith in God. Miracles are just not part of the scientific process.

Neither are miracles the only point of departure between the opposing realms of science and religion. Science involves a search for causes, it never prescribes life values; religion is a quest for meaning, wherein life values lie. Science is a "value-free enterprise, while religion is a valuational one." [2] Science explains, religion reveals, science informs, but religion reforms. [3]

Science involves careful observations, natural explanations, rational inferences, and deductive reasoning. Good experiments, quantifiable results, impartiality, and objectivity are desirable ingredients in good science. Experiments must be repeatable and verifiable, results should be reliable, and tests of validity are applied. And this entire process is called "science."

Science in Operation

Perhaps the easiest way to understand the scientific process is by doing a little science so we can see the operation of it as we go along.

First, we can make an observation. As Yogi Berra once put it, "You can observe a lot just by watching." Reputedly Isaac Newton saw an apple fall, and from his initial observation, he began a scientific study resulting in the formulation of Newton's law of gravity. Many laws and theories start with an observation. It is not a requirement that an observation come first, but frequently it does.

We may observe a pot of water on the stove. Once the burner is turned on, and a little time has passed, we might notice bubbles begin to form, and see the water turn to steam. We have all witnessed that phenomenon, and we could ask what causes the water to convert to steam.

One of our friends might suggest the fire god, Baal, is responsible, that the steam is an offering to him whenever there is water and fire. That might seem far-fetched to us, and therefore, an unsatisfactory explanation. So we could set up an experiment to see what really is going on. A simple experiment we might do is to place a thermometer in the pot, and see what temperature is necessary to boil water.

Before we commence, we might start out with a little conjecture, and guess that a temperature of 180 degrees Fahrenheit will do the trick. While we are picking out the pot, we send my little brother to the hardware store for a thermometer. Shortly he returns, hands over the thermometer, and goes out to play.

Now we do the experiment. We put our pot of cool water on the stove, stick in the thermometer, turn on the heat, and watch for the reading. Assuming for our simple experiment we are at sea level on a standard day, we notice the thermometer never gets above 212 degrees.

Not satisfied that one observation is enough, we repeat the experiment three times, and get the identical reading every time. Getting consistent results shows us that our test is reliable. Now satisfied, we record our findings, to wit: water boils at 212 degrees (at sea level on a standard day). We post a note on the refrigerator door and congratulate ourselves on our scientific achievement.

Next, your sister, who is a junior at the local state university, pops in to see what we are up to and admonishes us by pointing out that we may know what temperature it takes for water to boil, but we failed to discover why it happened. She proceeds to lecture us on BTU's and laws of heat transfer, the composition of the pot, the dew point of the air, the size of the room, the day of the week, and - "You didn't use distilled water. Don't you clowns know anything?"

This leaves us perplexed until my uncle, who is an engineer, and has overheard all of this, sits us down to offer a real explanation. Patiently he tells us what really happens: conductive heat causes the molecules of water to excite and vibrate at higher frequencies until they change form, and the water becomes steam.

The experiment we did was reliable in that it gave us consistent results, but it was not valid in that it failed to reveal the cause, which was what we were looking for after all. We should consult my uncle more often. Thank goodness someone knows the answers to questions like these.

There is a knock at the door. The next-door neighbor, who teaches science at the high school, heard all this through the open window, and cannot resist the opportunity to set us straight. She points out that although my uncle was right in a way, he too was merely making an observation.

Although we can see molecular action taking place, and we can observe heightened vibration with an increase in temperature, the reason substances change form, from solid to liquid to vapor and back again, at precise temperatures, depending on each particular substance, is a mystery to which no one really knows the answer.

After the teacher departs, we sit there bewildered. Outside the door, my little brother is telling his friends that water boils at 180 degrees because he heard us say so. My uncle is snoring on the couch, and the neighbor is watering her lawn. Dejected, we peel our notice off the refrigerator door. Well, so much for science, now who is this Baal guy our friend was talking about?

As ridiculous as that story is, it is intended to point out a bitter reality that has been played out by the actors and walk-ons that are parading across the stage in the creation-evolution theater. In the story, the friend attributed to Baal the conversion of water into steam. So have well-intentioned creationists touted

fact-defying explanations, invoking God at every juncture, supposedly to further the cause of Christianity.

Like my little brother in the story, pastors and lay ministers
have been vocal in their attacks on the evils of evolution without
bothering to check the facts. They take their cues from creationist
organizations, and assume blindly that they have done their home-
work.

Scientists on both sides of the issue, portrayed by the big sister,
lay on a barrage of superfluous evidences to build credibility into
their positions in an attempt to camouflage an otherwise incredible
explanation. This has been done by both creationists and evolutionists alike.

Also, there are venerated icons in this issue. Their expansive
knowledge and understanding holds us in awe. Like my uncle in
the illustration, they just make good sense to their followers, who
are swayed more by a powerful rhetorical argument than by the
logic of the conclusions.

In the role of the high school teacher, knowledgeable authors
have decried the paltry offerings from the extremists in the controversy. They have stripped away the veneer and exposed the fallacies on both sides of the argument, skewering creationists and evolutionists alike with equal fervor. Yet they have offered weak alternatives, leaving an awkward void in the middle. And, in case
you missed it, you and I took the roles of those frustrated and per-
turbed by the whole sorry mess.

Creation Science?

We know what science is, or is supposed to be. Let us see what
science is not. Creationists are quick to point out that evolution is
not a science. This is technically correct, since evolution itself is a
process, a tendered solution as to how the fossil record and the
genetic evidence can be explained. It may be correct to say that
evolution is a theory to which scientific methods can be applied,
but it is not a science.

Thus the term "evolutionary scientist" can be chuckled at, because a person who labels himself such has forsaken objectivity. By ascribing beforehand to a particular theory or rationalization, he disqualifies himself as an objective scientist. When one pins on a badge that says "Evolutionary Scientist," all impartiality is discarded. Impartiality and objectivity are desired components of the scientific method.

By the same token, the term "creation science" is another oxymoron - two words forced together which are normally contradictory, like "jumbo shrimp" or "British cuisine" (apologies to my English friends). Since "creation" is a proposed answer, and "science" involves an objective search for answers, the two words can not happily endure linked together. Also, since creation is a miraculous event with no known cause other than the act of a Creator, it falls outside the bounds of scientific testing.

Critics of so-called "scientific creationism" (a variation of the previously mentioned oxymoron) denounce invoking miracles to explain anything in the natural world. A widely held explanation for sickness was that God caused it as retribution for sin, until Louis Pasteur discovered bacteria. The "God caused it" explanation stops further inquiry, and here scientists have a valid complaint.

To scientists, creation will always be the explanation of last resort, after all other natural explanations have failed. Their attitude is: why invoke a miracle when it may not be justified, and go no further searching for causes? For believers, creation will always be the ultimate explanation, and it should be, but this says nothing as to what processes were involved, or what amount of time was used to bring about the creation. To some creationists, science is only a tool by which they can disprove all competing naturalistic explanations, leaving their own brand of creation as the only available answer.

As a result, the woeful image creationists have established to the scientific and academic communities is one of being utter nincompoops. It is precisely the misuse of scientific methods that projects an unsavory flavor to the entire creationist movement.

This is not to say that all creationists are guilty of the flagrant violations perpetrated by a vocal minority, but all get splattered with the same mud. For those involved actively in the scientific or academic communities, being labeled a creationist is one of the quickest ways to lose all credibility among peers.

In chapter 4, we will examine some of the infractions committed by a particular breed of creationists, and discover why all who believe in creation, and are vocal about it, are treated with such disdain. The point here is that creation and evolution have some things in common, and one of them is that neither is science. We might say that creation is true because we know it to be true intuitively, or because the Bible tells us so. There is nothing wrong with that, but call it intuition, or call it Bible teaching. Don't call it science.

NOTES

1 James L. Christian, *Philosophy: An Introduction to the Art of Wondering* (New York: Holt Rinehart and Winston, 1986), 243.
2 Holmes Rolston, III, *Science and Religion: A Critical Survey* (New York: Random House, 1987), 28.
3 Ibid., 28.

Chapter 3

AN EVOLVING QUESTION: MONKEY BUSINESS?

When words are bandied about like "creation" and "evolution," we find the words themselves can become so emotionally charged that their mere mention alone will arouse passions. The subject of human origins is a veritable mine field, where the mines have been set by scientists and clerics alike. Undaunted, we will tiptoe through the mine field of evolution, and hopefully, emerge with toes intact.

In an age when jet travel has become routine, and space missions are mildly anticipated events, we might think the subject of human origins should have been resolved long ago. Yet it is still the object of heated debate. For as many bridges of understanding man has erected, there are still chasms of ignorance left to cross. Emotions run high with this issue. Bias and prejudice overwhelm logic and reason, and as a result, many refuse to accept the evidence no matter how convincing it may be.

What is Evolution?

To ask the question "What is evolution?" can result in a variety of responses. Evolution is something everyone has heard about. Most people have formed opinions. Yet evolution is something few evangelicals have bothered to research. Evolution can be touted as fact, taught as theory, scrutinized as a working hypothesis, preached as philosophy, or propagated as a lie. Certainly evolution cannot be all of these at the same time - or can it?

While everyone tosses the word around, almost no one offers a definition. Critics and defenders alike attempt to avoid definitions because: (1) carelessly they assume their listeners already know what they are talking about; (2) it is a multi-faceted subject that resists one simple comprehensive definition; and (3) they do not want to show their ignorance on a subject in which they are supposed to be authorities.

Just agreeing on a definition can be troublesome. For starters, we can categorize evolution as theistic or atheistic, micro or macro, phenotypical or genotypical. Evolutionary theories such as Darwinism, Punctuated Equilibrium, and Catastrophism compete for prominence. Scientific theories of evolution are distinct from the philosophy of evolution, sometimes called evolutionism. Also, evolution can be broken down into component parts such as biochemical, biomorphical, biolineal, and biological.

It should be obvious that one catch-all definition will not work with this subject, although many people equate evolution with Darwinism. Dawkins defines Darwinism in his book, *The Blind Watchmaker*:

> In essence it amounts simply to the idea that non-random reproduction, where there is hereditary variation, has consequences that are far-reaching if there is time for them to be cumulative. [1]

Does that sound life threatening? Similar evolutionary phrases like "change through time" or "descent with modification" are fairly benign, are they not? After all, these defining terms do not disavow God's governance, or deny the possibility that a Creator certainly could work through evolutionary processes if He so desired, do they? In a later stage of his book Dawkins states its purpose:

> This isn't a student textbook. The basic idea of *The Blind Watchmaker* is that we don't need to postulate a designer in order to understand life, or anything else in the universe. [2]

Now here is where some would take issue. Normally a design necessitates a designer, and, quite frankly, we are rather well designed. Oh, we have aches, pains, maladies, sickness, disease, and birth defects, but we do not seem to be encumbered with any superfluous, non-functioning appendages. Overall, our bodies are fairly functional, and do we not have Someone to thank for that? But the perplexing, hackle-raising question is this: how did Dawkins journey from explaining a developmental process to denying an originating cause? The answer: He allowed an atheistic philosophy to infiltrate his otherwise scientific methodology.

George Gaylord Simpson illustrates how evolution has produced man ultimately, albeit accidentally.

> Man was certainly not the goal of evolution, which evidently had no goal. He was not planned, in an operation wholly planless. He is not the ultimate in a single constant trend toward higher things, in a history of life with innumerable trends, none of them constant, and some toward the lower rather than the higher. [3]

Futuyama joins the chorus:

> To the question "What purpose does this species serve? Why did God make tapeworms?" the answer is "To no purpose." Tapeworms were not put here to serve a purpose, nor were planets, nor plants, nor people. They came into existence not by design but by the action of impersonal natural laws. [4]

Here we can let Scripture answer directly: "And immediately the angel of the Lord smote him, because he gave not God the glory: and he was eaten of worms, and gave up the ghost" (Acts 12:23).

No designer? No goal? No purpose to life? Must evolution lead necessarily to agnosticism or atheism? Not necessarily, but usually it does. To be sure, there are theists who believe in a God-directed evolutionary process, but scant few of them are vocal. And those who do speak out need devices to waltz around uncooperative passages of Scripture.

Threat of Evolutionism

Reconciliation with God seems to be a human instinct that can be seen in early civilizations. Their "heart panteth," but their heads were fogged in by a belief in a pantheon, a hierarchy of gods. The object of worship was quite often the god of their city, who supposedly was in a direct line and subservient to a more distant supreme deity. Today, we see the opposite extreme. Rather than believing in a plurality of gods, it has been fashionable among the scientific intelligentsia to disdain any belief in God.

If evolutionary research involved nothing more than a quest for our origins using hard evidence, reliable data, and deductive reasoning, we could be totally unconcerned, but such is not the case. Regrettably, evolution has become a stepping stone toward a replacement religion - the "no God" religion.

The insidious philosophy of evolution leads unsuspecting followers into strictly human endeavors that essentially become religious. Issues such as human rights, animal rights, and environmental concerns reap the benefits. Of course, these crusades may have merit on their own. But when people hunger for a relationship with God, and their need is derailed because they have been misled to believe the Bible is unreliable (contrary to the "truth" of evolution) they may obtain a measure of satisfaction through a golden idol, a "worthy cause."

In denying the existence of a Grantor of eternal life, evolutionary philosophy denies the hope of eternal life. If we do not live forever, if we are only a step on the evolutionary ladder, then let us busy ourselves making our lives a little better, and the lives of our successors. The philosophy of evolution creates a void, and then seeks to fill that void, such that recycling of scarce resources, or preserving the environment, or saving precious animal life is thrust into the vacuum. The sweet wine supplants the water of life.

What Christians should be alarmed about concerning evolution is not where man came from, but where is man going? There is fierce competition for the hearts and minds of men. That is nothing new. Only the methods are revised from time to time. And the philosophy of evolution has become one more weapon in the arsenal of the Lord's enemies.

Defining Evolution

So what is evolution all about? In *The Monkey Puzzle*, authors Gribbin and Cherfas talk about the "fact" of evolution:

> Since the time of Charles Darwin, ... it has become a well-established fact that we are descended from monkey-like ancestors, tree dwellers who thrived in the tropical forests of 35 million years ago. [5]

Not everyone agrees man's descent from monkey-like ancestors has become so universally accepted that it has reached the lofty designation of "fact." Most would still put it in the theory stage. And yet, is evolution a fact? In a word, yes. If by the word "evolution" we mean to imply nothing more than "change through time," or "genetic shift," or "descent with modification," then we can concede an evolutionary fact that is pertinent to the overall theory of evolution.

It is a fact that if we view the upper part of the geologic column containing fossils, going from bottom to top, the trend is from simple to complex. With the exception of some mountainous areas where folding has occurred due to active plate tectonics, we can say that, in general, complex fossils overlie simple ones. We can call that *biomorphic evolution.*

C. S. Lewis said "all powerful lies" were "based on a truth."[6] Evolutionism (and, sadly, creationism, too) starts with a fact, which can be the foundation to support divergence and digressions. Whatever might have been true at the inception, can be contaminated by extrapolation. Evolutionists do this by parlaying on the revelations of biology and genetics, and using their data-based wisdom, they deny the creation, the Creator, and His Word.

A step down from the fact of evolution is the theory of evolution which is based on the geologic column being viewed as a record laid down in time. The fossil record supports the concept that simple organisms begat increasingly complex types. This descent through time with movement toward complexity is called *biolineal evolution.*

One step removed from the theory is the working hypothesis of evolution. The hypothesis is that all living creatures are but branches on the phyletic tree leading from one common trunk. In other words, so goes the hypothesis: humans, animals, and even plants have common ancestry, everything that lives is related.

Supposedly, every living creature is biologically and genetically linked to every other living creature to include all extinct creatures as well. This is *biological evolution.* Enough evidence might dispel doubts about a link between hominids and other primates, but that would not prove necessarily that we humans are distant kin to horseshoe crabs.

Creationists have offered alternate explanations to biological evolution allowing a certain amount of "microevolution," or limited evolutionary processes, within parameters; the Bible says "kinds" (Gen. 1:21). No one seems to know exactly what constitutes a "kind," and so the issue is clouded a bit. There is little evidence to disprove biological evolution, but it has been argued that not enough supporting data has been accumulated to lift the entire concept of biological evolution above the level of working hypothesis. The term "theory of evolution" or "evolution theory" is generally accepted, so using this nomenclature will lessen confusion, but a term such as the hypothesis of evolution might be more precise, due to incomplete data spanning over 3 billion years of life on earth.

Conclusive proof for biological evolution has not been discovered, but the science of genetics offers the best chance of finding it. And the results of current research are generally supportive. But what eventually comes out of the science of genetics may be inconclusive. The verdict is not in yet.

Continuing down the scale we come to conjecture. A good example of that is *biochemical evolution,* the notion that life in some way came from non-life. A particular batch of chemicals in a "primordial soup" got lucky somehow, and arranged itself into an organism complete with its own DNA. Figuring out how to replicate was no easy task, but once this was accomplished the sky was the limit. Increasing complexity followed as night follows day; with enough time, and enough fortuitous accidents, here we are!

To be sure, there are variations on the theme. At deep sea thermal vents, organic compounds may have descended on clay or silicate templates and somehow turned into "living" replicators using RNA, bypassing the DNA step completely, starting the ladder of life at the bottom rung.

No matter what manner of explanation is used, the difficulty with life popping itself into existence from non-living chemicals, or even from naturally occurring organic compounds, is that even at the level of the microbe, life is incredibly complex. Living organisms have such intricate, interconnected inner workings that the sheer probability of life emerging from a chemical soup is so near zero as to be for all practical purposes - zero.

The last digression is the fallacy (some would say the lie) of evolution. Here, evolution is deified. Philosophy dominates, and phrases like "grand evolutionary synthesis" take on the form of a religion. Not only complex creatures, but complex social structures, too, are seen to flourish from purely evolutionary processes.[7]

Spurning a Creator-Designer God, atheistic evolutionists create a god of their own design - a god that makes no demands, exacts no tribute, and, above all, does not judge. What should be obvious is that a non-God explanation is exalted by many evolutionists primarily because the alternative is unthinkable. If there is really a God then - we may be accountable! Who wants that?

A belief in no God is as much a religious entrapment to the nonbeliever as trust in a Creator-God is a constraint for theists. A reliance on accidental processes as the supreme deity exacts an even greater measure of blind faith from nonbelievers than is required from those who profess God. There is no supporting data. Atheists who wrap themselves in science are like emperors with no clothes. Science demands evidence to support theories and hypotheses, and no evidence has been found to justify a purely accidental scenario.

Darwinism

Fraudulent science deserves criticism, but what about honest science? Berating Darwin has become fashionable, coming from Christian writers and evolutionists alike. Where Darwinism has been put forth as a religion, it needs to be denounced, but we need to reflect a little before we offer up Charles Darwin to public scorn.

Darwin was reared in the faith. Ironically, his voyage aboard the HMS Beagle was dedicated to confirming the truth of the Bible. Could a devout young naturalist kneel before his God and ask His blessing for such a trip? And if God revealed His truths in nature to a young Darwin, can we ridicule this scientist for reaching the conclusions he did? As Ruse put it:

> There was no barrier posed by religion for Darwin, because ultimately Darwin did not see religion and evolution in conflict! [8]

Darwin was familiar with pigeon breeding and knew that varieties of pigeons could be produced by selective breeding. In the wild, nature does the selecting, and he called that "natural selection." Should we take offense? Nature is in God's domain, is it not?

The work of Gregor Mendel (1822-1884) was ignored until after his death. Mendelian genetics explained how genes (he called them factors) produced variant offspring, but his work was unknown in Darwin's day, and, of course, no one had a clue about DNA. Still, with what was known about heredity, Darwin ventured that the "raw stuff" of evolution, whatever it turned out to be, was unplanned or "random." Coupled with natural selection, all of the present-day plant and animal life could have derived from common stock. [9] All of life, as Darwin saw it, shared common ancestors.

To be sure, Darwin was unaware of such things as plate tectonics, or periodic catastrophes that have wiped out animal life *en masse* from time to time, or the mechanics of DNA replication. Had he known what we know today, he might have formulated a slightly different theory. But for what was known prior to 1859, when *The Origin of Species* was first published, Darwin did a laudable work. God allowed him to discover the unvarnished truth, and Darwin reported his observations and conclusions honestly, which is all we can expect from a working scientist.

Isaac Newton's mathematical formula pertaining to the law of gravity was revised after the acceptance of Einstein's theory of relativity. No newspaper headlines blared, "Newton's Law of Gravity Repealed!" His formula was simply changed to reflect that space-time was curved, not linear as Newton had thought. Darwin deserves similar treatment. A recent book has been published called *Darwin on Trial*. What could poor Charles possibly have done to be hauled into court? Let's be sensible. Darwin's initial theory needs updating and refining with the benefit of added knowledge.

In opposition to Darwin's gradualism, punctuated equilibrium, first posited by Niles Eldredge and Stephen Jay Gould, is gaining support. According to this model, millions of years of relatively small changes in the characteristics of populations have been punctuated from time to time by abrupt large changes within only a few thousand years. Eldredge and Gould observed species that had gone unchanged for millions of years suddenly gave rise to new ones. [10]

As interesting as the evolution argument may be, as varying models compete for prominence, it should pose no threat to Christian faith as long as discussions remain entirely within the realm of science. Let honest scientists set about the task without atheistic philosophy intruding and without harassment from ill-informed Christians! A revision in the light of what we know today certainly can be undertaken without digging Darwin's bones out of Westminster Abbey, and dumping them in a pauper's grave.

Dabbling in DNA?

Each person's genetic blueprint is contained in his or her DNA. Every cell in our body contains our unique genetic make-up. The approximately seven foot long strand of human DNA found in each human cell has about 3 billion nucleotide bases. Through the process of nucleotide substitution, subtle differences are passed on to succeeding generations.

Evolutionists attribute these substitutions to chance variations, and therefore, do not invoke any mechanism for improvement. Differences add up in time, and thus, organisms change. Natural selection decides which mutations are passed to future generations.

As the theory goes, a nucleotide substitution might result in a large and bulbous nose, for example. A person sporting such a protrusion might find it difficult to attract a helpmate. That person might remain single and pass from life's scene leaving no bulbous-nosed progeny - natural selection at work, but this does not rule out a Creator-Designer God.

In Darwin's theory, the word "random" causes most of the consternation. Predominately, it is the prospect of a random process in operation that has caused atheistic evolutionists to rub their hands in glee, and Bible believers to go ballistic. Primarily, this is due to equating the word "random" with "accidental" or "unforeseen."

In Templeton and Herrmann's *The God Who Would Be Known*, the scriptural basis for "chance" is underscored:

> Chance in the neutral scientific sense, however, is mentioned as a part of God's plan. "The lot is cast into the lap," says Proverbs 16:33, "but the decision is wholly from the Lord." Here is a clear indication that God is the Lord of events, which in this sense happen by chance, just as much as he is Lord of those that seem orderly to us. It may be easier for us to see God's hand in the obviously orderly pattern; but the Bible seems to exclude the idea that he must always work in this way. The either/or notion (either God or chance) is simply not the way the Bible relates the two ... [11]

Randomness does not mandate lack of direction. Again, this is the same confusion about evolution that we find in Christian doctrine concerning the seeming paradox of free will and predestination, discussed previously. God sees the end from the beginning.

The concept of random nucleotide substitutions should be no more threatening to Christian doctrine than free will.

We make virtually thousands of random decisions every day. We blink our eyes, draw our breaths, take our steps, twiddle our thumbs without giving these things any thought whatsoever. There are billions of eyes blinking away, totally at random all the time. There is no more necessity for God to control eye blinking than there is for Him to control nucleotide substitutions on the DNA molecule. Due to God's omniscience, He simply knows - that is sufficient.

Most of us have used video tape recorders. When we are away from home we may tape sporting events, football for example, so we can watch them later. Not knowing the outcome, we enjoy the games on videotape with similar excitement as if they were live.

In life, God knows the final score while we are watching the game live and it is still in the second quarter. He already knows the passes that will be dropped, caught, or intercepted, the quarterback sacks, the blocked kicks. His foreknowledge permits prophecy. In times past He has told us about future events. Some of those events have happened while others are to be fulfilled.

Even random nucleotide substitutions that cause genetic mutations with accompanying speciation and diversity of life do not preclude God's foreknowing the resulting life forms. Just because the outcome is known does not mean God must control the outcome to preempt our free will, or to preclude random processes. In Arthur Peacocke's book, *Creation and the World of Science*, he concludes:

> I see no reason why God should not allow the potentialities of his universe to be developed in all their ramifications through the operation of random events;... [12]

Nothing in Scripture prohibits God from using special creation at the inception of the lowest microbe, and permitting a process of descent with modification to culminate even in modern man.

Confirming a Creator

If the theory of evolution from the Big Bang to modern man is to be believed in its entirety, it would need to have a plausible explanation for contrary evidence. There are two distinct areas in which the commonly advocated theory of evolution or "Darwinism" fails to offer satisfactory answers.

The first is in the area of Deity. What we see in nature itself is not the only means we have to conclude that ultimately a benevolent Creator-God is responsible for life as we know it. Clearly the empirical evidence of design implies a designer, but all the eggs are not in one basket.

There are at least three other independent sources by which we should be able to confirm God's existence easily without casting a glance at nature. God's indelible stamp can also be found (1) in history; (2) in fulfilled prophecy; and (3) in personal testimony.

The Bible has proven to be an invaluable source to archaeologists searching for ancient cities. We will see in later chapters that contemporary literature in early Bible times reinforces the inspired text.

The prophet Isaiah laid a prophecy on the wicked city of Tyre that was carried out by Alexander the Great to the letter, hundreds of years later. Jesus warned of the destruction of Jerusalem, "... the days will come, in the which there shall not be left one stone upon another, that shall not be thrown down" (Luke 21:5, 6).

For a heart-rending account of what God allowed to fall upon the Jews in fulfillment of prophecy, read the chapter on "The Destruction of Jerusalem" in Keller's book, *The Bible As History*.[13] Ironically, the destruction of Jerusalem, prophesied by Christ, was witnessed and recorded by the Jewish historian Josephus - a non-believer.

Men have gone to prison, swung from gallows, stood before firing squads, and have been electrocuted in electric chairs solely on the basis of personal testimony. Our legal system holds sworn testimony in such regard that permanent incarceration and very permanent physical death can result. In some cases, only one eye witness was sufficient.

Millions upon millions of believers could testify to a living God who has acted personally in their lives. An unending procession of witnesses could come forth and offer testimony. There is abundant proof of God operating in life in general, and in our own lives specifically, to quash all objections. We may not be able to prove "scientifically" that this is the God who created heaven and earth, but the Bible declares that He did. And there is no better explanation.

In the absence of data to the contrary, even pragmatic scientists could acknowledge the Creator, if only tentatively. Yet, there are those who wear the label "Scientist," who offer incredibly flimsy alternative explanations, and their disciples also number in the millions.

The second area in which the theory of evolution fails to give satisfactory answers is imbedded in and inseparable from the very data Darwinists have collected to support their theory of descent with modification. What has been overlooked is an obvious hypothesis, dare I say theory, suggested by the data.

What has escaped general recognition is that organisms have demonstrated a capacity or a capability to make genetic alterations to accommodate changes in the environment. Although natural selection might explain how adaptations can be passed down from generation to generation, giving subsequent populations an edge in survival, it does not resolve how adaptive mutations arose in the first place.

Evolutionists will say that within the gene pool of populations there is enough genetic variability to allow for selections to be made, which can give succeeding generations an advantage. The adaptive characteristic is expanded as it is passed down to privileged offspring. But this is an insufficient explanation that ignores what we see in nature.

Organisms do appear to make unique and very specific changes in order to better equip themselves to survive difficulties encountered from environmental shifts. Some could not make accommodations and became extinct, but those that did survived. Specific adaptations have occurred in populations that have been exposed to an environmental threat, and lived through it. Sufficient evidence exists in nature to support this hypothesis: Organisms seem to be endowed with an adaptive quality enabling them to make specific genetic mutations when the need arises, and those changes may be inherited.

Furthermore, the opposite seems to be true. Organisms tend to shed useless features when the need for them ceases. In other words, there is empirical, supporting evidence that a type of stimulus-response mechanism can induce adaptive changes, that may be incorporated in the DNA of organisms, and passed to future generations.

In addition to environmental factors being an instrument for modification, usage also may play a part. Human beings developed larger brains than their predecessors because they used them. Increased thinking capacity may accrue to the children of thinking parents. And the same thing may be true of other physical attributes.

Antibiotics experts have long recognized this adaptive capacity in microbes that have developed immunities to the drugs designed to wipe them out. Penicillin was introduced in the early 1940's. Soon after the infectious disease-causing bacteria were exposed to penicillin, they began producing an enzyme called beta-lactamase, which destroys penicillin and related antibiotics.

In the early 1980's broad-spectrum beta-lactams were launched to kill drug-resistant bacteria. But the bacteria responded by mutating the gene encoding its defensive enzyme so that it now can ward off these drugs too. George Jacoby, a specialist in infectious diseases at Massachusetts General Hospital, remarked, "Bugs are always figuring out ways to get around the antibiotics we throw at them. They adapt and come roaring back." [14]

Researchers also know that certain genes have a DNA repair function. They even know there are several DNA repair pathways. Some genes are capable of repairing DNA without making error, while other genes, in their words, "are prone to make mistakes." [15] It is suspected strongly that these latter genes that repair DNA with new coded information cause mutations which contribute to evolution.

Recent outbreaks of tuberculosis have fostered new research into this disease. It has been demonstrated that due to a mutation, the bacterium causing TB is now resistant to isoniazid, the main drug used in treatment. Thanks to a hard-working research community, a genetic basis has been identified for TB drug resistance. [16]

What has been recognized already at the level of the microbe can also be witnessed at the highest levels of life. Among human populations, skin color affects the absorption of vitamin D from sunlight. Higher latitudes have decreased sunlight, lighter skin improves absorption, and lighter-skinned peoples are found at higher latitudes.

Sickle cell anemia is a genetic disease affecting some black populations. This gene is recessive and appears to afford enhanced resistance to malaria. The sickle cell trait may have been a genetic response to an environmental danger.

There is an increased risk of inheriting the genetic disorder, Tay-Sachs syndrome, among Ashkenazi Jews, and this has been traced to Polish ghettoes in World War II. Though, like sickle cell, the disease is fatal where one inherits the gene from both parents, yet the Tay-Sachs gene has been correlated to an increased resistance to tuberculosis, the scourge of the ghettoes in those days.

On the other side of the coin, there have been found in caverns fish, crayfish, and beetles that no longer have vision. Indentations still remain where eyes once scrutinized the world millions of years ago. Through countless succeeding generations, born in total darkness, eyes were jettisoned by these creatures as needless encumbrances.

Researchers have begun preliminary investigations in this general area of inheritance affected by environmental factors. A conference was held in Pittsburgh in September, 1992, on "male-mediated toxicity."

> After 3 days, the consensus was that there is an urgent need for studies to elucidate mechanisms underlying tantalizing evidence that many different types of paternal exposure induce changes in sperm or semen that could affect children's health. [17]

In the not too distant future we may discover how adaptive genetic mutations may be shaped by environmental forces, something that was postulated by the earliest pioneers of evolution theory, J. B. Lamarck and Charles Darwin. [18]

Genetic adaptations made by microscopic organisms, insects, animals, and humans point in the direction of change through time through genetic mutations which could be affected by usage or the environment, and are inherent as part of our immune-response system. These factors may induce the production of enzymes triggering genes that modify the DNA. The modified DNA may cause beneficial adaptations for succeeding generations. And who might have had the foresight to endow His creatures with such a beneficial characteristic? How about a benevolent Creator-God?

The explanation supported by some of our most eminent biologists today, however, is that chance mutations occur spontaneously and accidentally in populations affording enhanced survivability to the lucky descendants that inherit those beneficial adaptations. But consider: Which answer better fits the evidence?

Adapting

How would God give organisms adaptability? There are at least two possibilities. Maybe God intervened directly, making the necessary adaptations on an "as required" basis when a particular need or threat arose or subsided. Or, perhaps God built into his creatures the inherent capability to make necessary adjustments which could then be passed to succeeding generations.

Either a "hands on" or a "hands off" approach could account for adaptive accommodations, but the "hands off" method looks to be superior. In the case of sightless creatures found in caves, it is unlikely that God took sovereign action to cause them to shed properties no longer needed.

Although Darwin predicted that need-generated accommodations could be inherited, this concept has been discarded by Darwinists today. To every instance of environmental adaptation, the

Darwinist maintains the fortunate organism was only lucky it already possessed within its DNA a specific sequence which became expressed as an adaptive quality for enhanced survival through selection. Perchance we are lucky, but perhaps, we are blessed.

An explanation that better fits the evidence is that a force of direction was exercised either at the inception or throughout. Through a process we are only beginning to uncover, an incredible intelligence, the evidence supports, wrought His creation through time to culminate in man. And He did it without directly controlling or manipulating the mechanism.

Creation to Degeneration

Knowing that a Creator started the life process could lead us to ask whether mutations in a gene sequence resulting in new species, including man, should be considered creation events, or are these punctuations in life natural occurrences resulting from random processes? For an answer, we need to look at the results. Most gene mutations in a population have a detrimental effect. Researchers have identified thousands of diseases in humans resulting from mutations. If the genetic changes in the past had been as deleterious as we see today, it would be hard to imagine species development moving in a positive direction. But, does that imply guidance?

Should we credit at least the direction evolution has taken to the guidance or governance of a Creator? Perhaps not. If God was directly responsible for the positive mutations that have occurred in eons of time to bring about this marvelous creature we call man, then to whom do we attribute these apparent genetic blunders resulting in genetic disorders? Credit and responsibility go hand in hand, and therefore, lest we wish to accuse the Creator, it seems unlikely that God manipulates DNA, either for better or worse. Random processes are a viable explanation, yet a significant change apparently has taken place at some time.

Was there a point in time when the gears changed from a gradual refinement of the genetic code to a different set of rules? The evidence suggests there may have been. The cessation of perfecting and improving seems to be commensurate with God's bringing to a halt His creation activities on the sixth day.

A shift toward degeneration may have started with His seventh day of rest. Is there evidence that a shift from genetic improvement to genetic degeneration actually happened, and can we pinpoint it in time? That would be a tall order, and beyond the scope of this discussion, but there are clues.

The geologic column reveals that new species emerged in profusion with increased complexity during the past. Yet, in just the last few hundred years, we have observed few if any new naturally occurring species, though thousands become extinct every year. This change in direction from improvement to degeneration may have started after the creation of Adam and Eve, or it might be seen as a consequence of the Fall.

Common Ancestors?

The fossil record is as open to interpretation as the Genesis record. There is much ballyhoo about the lack of transitional fossils at branch points. For example, it is commonly believed that mammals branched off from reptiles in the Permian period about 280 million years ago. A group of reptiles called synapsids stand in the breach. These are classified as "reptile-mammals" due to their possession of characteristics common to both. These first appeared in the Carboniferous, flourished in the Permian, and became extinct in the Jurassic. [19]

Classifying these creatures as intermediates has been challenged by some who maintain that most reptiles have three jawbones and one ear bone, while mammals have three ear bones and one jawbone. According to this complaint, intermediate animals should have these bones somewhere in between. So, whether transitional life forms existed or not is driven as much by presupposition as it is by the revelations of the fossil record.

What we can say is that transitional forms of life were less likely to become fossils than were well-established, far-ranging animals. Smaller populations in geographic isolation produce morphological changes faster in a given period of time than does a larger population scattered over a greater geographical area. Change is more rapid in a small gene pool. Therefore, transitional life forms would have been less likely to become fossils due to their relative scarcity. It is questionable, though, whether this alone can account for the scant evidence of intermediates.

The impetus for postulating "gaps" in the fossil record has been to substantiate that if there are real gaps between animal species, this supports special creation either at the beginning or throughout. God would have either created "kinds" at the inception, or else He introduced unique kinds periodically on an as-required basis.

This gap methodology tries to establish that if such gaps could lie between lesser members of the animal kingdom, then there may also be gaps, and therefore no ancestral relationship, between other higher primates and hominids, or between early hominid prototypes and *Homo sapiens*, or between modern-looking humans and those that came from Adam.

Gap mentality overlooks something basic. It is much easier to examine and make reasonable inferences from the last four million years which includes mankind than it is to decipher what was going on 280 million years ago when mammals began, for example.

Just as scientists should be willing to accept a Creator as responsible for starting life and ordering its processes when the data supports that, so should Bible believers acquiesce when the empirical evidence bolsters the case for evolution in the absence of contrary data or biblical restraint.

Even though many Christians endorse evolutionary concepts, some will say they cannot accept the notion "we came from monkeys." Even though biologists agree we did not come from monkeys, making such a statement of adamant denial speaks volumes about one's closed-mindedness. Using the words "from monkeys" reveals a commonly held misunderstanding. My brother and I share common ancestry. I did not "come from" my brother, nor he from me. Although certain animals may be related, this does not mean necessarily that one is directly ancestral to the other.

It is a widely held theory that man shares common ancestors with other higher primates. Specifically, biologists consider man and chimpanzee to be on the same branch of the phyletic tree with the gorilla branching off earlier. This does not mean necessarily that any one of the three descended from the others.

Even though many also disagree with the concept of shared ancestry, at least it states the theory more accurately. If the question was simply, did man come from monkeys, the short answer is that it appears man did not. Life, however, snubs simplicity.

A Creationist Counter-Argument

An examination of the human DNA molecule, the genetic blueprint of man, shows that many of the same sorts of genes and gene sequences are found in lower animals. Some creationists would say this is not surprising. Since man has many of the same physiological functions as lower animals, our DNA naturally would carry comparable information.

Thus the argument has been made that any resemblances in DNA mean only that God used similar gene sequences to order similar functions. Although there may be a commonality of design, as the argument goes, that does not prove common descent.

A typical theological model of the origins of *Homo sapiens* mandates a species created uniquely with no ancestral ties to any other species. According to this theologically oriented hypothesis (or conjecture), similarities exist due to the Creator's having modeled our particular DNA molecule after other less noble creatures.

Man's DNA sequence, it is argued, is an example of a common design used by God, who created all animals, and thus He simply used like patterns of DNA expression. That might seem logical if we were to subscribe to intermittent creation, and if our DNA contained only functioning genes.

Common DNA Markers

The case for common design but no common descent becomes suspect when the entire human DNA sequence is analyzed, and copying errors are found in the same places in the DNA of non-humans. In addition to genes that function normally, we have nonfunctioning genes as well, called "pseudogenes." Our DNA sequence is a complicated set of instructions that appears to have a long history of replications, and therefore contains an abundance of pseudogenes. Humans have pseudogenes incorporated along the entire DNA sequence that can also be found in other animals, i.e., the chimpanzee and gorilla.

It is one thing to suggest that God may have modeled our DNA along the same lines as lower animals, such that similarities are due to like genes ordering a protein sequence serving a like function. It is quite another to assert that God also incorporated all the excess nonfunctioning baggage too. Why use the same trash, accumulated through time, inherent in the DNA sequence in lower animals to model the DNA sequence in a supposedly unique creation - man?

We would be left to question what kind of creator would preserve the functioning DNA without purging the nonfunctioning elements. The answer would be: a creator who wanted us to believe that man was genetically linked to the lower animals, but was not actually so linked. Would God implant false evidence to lead us to erroneous conclusions? If so, to what purpose?

Further evidence of genetic linkage between man and other higher primates can be derived from an endogenous retroviral sequence imbedded in our DNA that is also found at the same point in the DNA of chimpanzees. [20] Retroviruses are a particular class of virus, which includes the HIV virus that causes AIDS, for example. These viral agents have the ability to annex themselves directly into a DNA sequence, and an ancient virus apparently did.

The entire genetic code was then passed to future generations, including the retroviral sequence. This retroviral sequence has no activator mechanism, and thus is harmless, but here is the point. Not only do man and chimpanzee have the same number of muscles, bones and teeth, our DNA has a correlation of nearly 99%. [21] But in addition, an identical alien viral sequence can be found at the same locus point on both human and chimp DNA.

A rational explanation is that the viral sequence became attached to the DNA of a common precursor. It has remained in the DNA, and has been copied in both man and chimp for millions of years. This additional confirming data for relatedness to an animal that already looks to be a close relative anyway, makes a strong case for a brachiating forbear on our family tree. The theory that the highest primates, including man, are all divergent twigs off a common branch has strong supporting evidence. Whether we like it or not, gorillas, chimps, and man do not appear to be separately created entities.

The proof of genetic linkage between man and other primates may not be absolutely ironclad, but it is abundant, persuasive, and maybe even overwhelming. At least, the theory of genetic linkage, or shared common ancestry, fits the empirical evidence. Furthermore, genetic similarities have allowed us to do research on lower animals with direct application to humans. Mice have been imbued with human genes to produce human antibodies. The Rhesus monkey helped us uncover the Rh factor, an important bit of information that has saved thousands of newborn children. Chimpanzees are used for AIDS research. Genetic closeness has been of tremendous value, and may prove to be even more so in the future.

Our federal government has spent billions of dollars so far to map the human genome. Although still incomplete one thing is sure: some non-human genomes will be mapped before they are done. Soon we may have additional compelling evidence of biological linkage between humans and non-humans; some geneticists think we have it now. [22]

The burden of proof really falls on those who advocate no such linkage and deny shared ancestry to come up with a model that would explain the data better, or at least in equally satisfying terms. To date no such model has come forth.

If we examine the evidence fairly, it appears that we may have cousins, some would say brothers, who eat termites and live in the forest. If this disturbs us, we can take solace in that, were they to find out about it, they might be similarly offended.

If it turns out that lesser creatures are part of our heritage, then where do Adam and Eve fit in? The short answer is that they fit in about 5000 BC. The long answer is what this entire book is about. We will see that ancient and shared ancestry, even if confirmed beyond doubt, will not preclude an entirely trustworthy Bible that can be taken literally - even Genesis.

NOTES

1 Richard Dawkins, *The Blind Watchmaker* (New York: W. W. Norton & Company, 1987), xi.

2 Ibid., 147.

3 George Gaylord Simpson, *The Meaning of Evolution* (New Haven: Yale University Press, 1967), 293.

4 Douglas J. Futuyama, *Science on Trial: The Case for Evolution* (New York: Pantheon Books, 1983), 37.

5 John Gribbin and Jeremy Cherfas, *The Monkey Puzzle* (New York: Pantheon Books, 1982), 13.

6 C. S. Lewis, *Mere Christianity* (New York: Macmillan Publishing Company, 1952), 78.

7 Ervin Laszlo, *Evolution: The Grand Synthesis* (Boston: New Science Library, 1987), 87-109.

8 Michael Ruse, *Darwinism Defended* (Reading: Addison-Wesley Publishing Company, 1982), 26.

9 Ibid., 27.

10 Richard A. Kerr, "Did Darwin Get It All Right," *Science* (10 March 1995), 1421.

11 John Templeton and Robert L. Herrmann, *The God Who Would Be Known* (San Francisco: Harper & Row, Publishers, 1989), 58.

12 Arthur R. Peacocke, *Creation and the World of Science* (Oxford: Clarendon Press, 1979), 94-95.

13 Werner Keller, *The Bible As History* (New York: William Morrow and Company, Inc., 1981), 364-373.

14 Ann Gibbons, "Exploring New Strategies to Fight Drug-Resistant Microbes," *Science* (21 August 1992), 1037.

[15] G. Strobel, "Mutated plant sheds light on DNA repair," *Science News*, Vol. 144, No. 13, 198.

[16] Peter Aldhous, "Genetic Basis Found for Resistance to TB Drug," *Science* (21 August 1992), 1038.

[17] Richard Stone, "Can a Father's Exposure Lead To Illness in his Children?" *Science* (2 October 1992), 31.

[18] Ruse, *Darwinism Defended*, 32.

[19] P. Arduini, and G. Teruzzi, *Prehistoric Atlas* (London: Macdonald & Co. (Publishers) Ltd., 1982), 56.

[20] T. I. Bonner, C. O'Connell and M. Cohen, "Cloned Endogenous Retroviral Sequences from Human DNA," *Proceedings of the National Academy of Sciences* (1982), 79: 4709-4713.

[21] Gribbin and Cherfas, *The Monkey Puzzle*, 15.

[22] Edward E. Max, "Plagiarized Errors and Molecular Genetics: Another Argument in the Evolution-Creation Controversy," *Creation/Evolution* (XIX, 1986), 34-46.

Chapter 4

YOUNG-EARTH CREATIONISM: A LITERAL ENIGMA

Alexandria, Virginia, a suburb of Washington, D.C., was founded in 1749. Thomas Lord Fairfax, and his cousin, Colonel William Fairfax, advocated building this town on the "Potowmack River" and petitioned the Virginia governor's office at Williamsburg. [1]

After approval, the port site was planned and its streets were surveyed. A certain 17 year-old assisted the county officials in the survey work. [2] The eldest son of Mary Ball Washington, George would go on to become commanding general in the Continental Army, and served eventually as the first president of our fledgling nation.

Hessian soldiers laid the cobblestone streets of Alexandria, according to legend. One can still see a block of Princess Street, between Washington and St. Asaph streets, as it looked in bygone days. Actually, this street was resurfaced in 1979, but "using original cobbles" according to the bronze plaque in the middle of the street. Pedestrians still stroll along the brick sidewalks of Old Town Alexandria, weathered, worn, and wavy, though some have been replaced over the years.

Local builders, building inspectors, and knowledgeable real estate agents can enumerate interesting features to be found inside these quaint old townhomes. Of course, refinished hardwood floors and sparkling new kitchens complete with garbage disposals, ice-maker refrigerators, trash compactors, and microwave ovens also abound. After all, historic is one thing, primitive is another.

Clapboard townhomes with functional wooden shutters still adorn St. Asaph Street, some with lead-based paint peeling away, and some restored and gleaming. For one who would like to reside on this historic corner, but who does not discriminate between owning an old home versus one that just looks that way, two new (but old looking) "Georgian" townhomes have been built recently.

Indeed, all of Old Town Alexandria is a veritable patchwork quilt of genuine old homes interspersed with modern houses built over the years to what could be described as arbitrary historical specifications. If someone was racing between stop signs on St. Asaph Street, he might not notice which homes are truly historic, and which only appear to be. But if we walked along we would detect the differences easily.

If copper gutters and down spouts, tin and slate roofing, distorted glass panels, and noticeable weathering are not enough, the Historical Alexandria Foundation furnishes registered bronze plaques that can be brandished conspicuously on the homes that are truly historic. Also, the courthouse records will certify the age and history for anyone seeking the information. The point being: it would be difficult to make a mistake on the age of any of these homes if one is willing to spend a little time, seek competent counsel, pay some money, or do a bit of research.

The same is true in estimating the age of the universe and the earth and its contents. No one could make a mistake if he bothered to exert even minimal effort. Astronomers and astrophysicists attest to an ancient universe on the order of 12 billion years or so. Geologists certify an antique earth of about 4.6 billion years. Biologists have traced the beginnings of life to over 3 billion years. And the hominid line has a long history documented by fossils spanning a few million years, according to paleontologists and archaeologists who make their living knowing such things.

Pastors, evangelists, and lay persons who are not steeped in science can ignore the data, and expound on all areas of Christian doctrine, including God's creation described in Genesis. There is no harm in that. Disclaiming expertise in science should not impinge on one's ability as a Bible expositor. But a potential for great harm exists when religious leaders with no expertise in science also make no investigation, and expound on scientific matters, possibly misleading their flocks. (See for reference Matt. 15:14.)

Creation in Court

The Constitution protects our freedom of religion right, giving latitude to teach almost anything but insurrection in the pulpit and in our private Christian schools. Public schools must satisfy a broad audience, are answerable to local boards of education, and have proved to be resistant to creationist explanations.

In a 1988 seven-to-two decision, the Supreme Court struck down a Louisiana statute that would have required equal time treatment for the teaching of "creation science" whenever the theory of evolution was taught. The high court ruled that states may not mandate teaching creation in public schools.

That decision labeled creationism and creation science a religious belief, rendering it unsuitable as a required classroom subject. Although the court ruled properly, in my estimation, it was a classic example of a right result for all the wrong reasons. Creation science should be banned from the classroom not because it's religion, or because it's bad science, but because it's bad religion.

Although the last legal brief has yet to be filed in the creation-evolution controversy, a certain amount of perspective would be helpful. We do not live in an atheistic country. American coins, even those resting in the pockets of Supreme Court justices, still say, "In God We Trust." Our Pledge of Allegiance reminds us that this nation is "under God."

Sessions of Congress are opened in prayer. By tradition every American president since Washington has laid his left hand on the Bible when taking the oath of office. Belief in God or a "Higher Power" is not an issue to most Americans. For the most part we believe in a Creator-God.

Creation Timing

As to the manner in which this universe, life, and man came about, the question for theists is: Did He snap His fingers, or did He twiddle His thumbs? The major sticking point between creationist solutions lies not with who did it and why, but rather, how and when.

Christian fundamentalists are in total agreement on who was responsible, affirming that heavens and earth and life were created by God. In the Judeo-Christian tradition, God is a personal, caring, loving father who has numbered the hairs of our head. According to evolutionists, we could have had two heads as easily as one. So fundamentalists do have a legitimate complaint.

Creationists (who may or may not be fundamentalists) start from the same premise that God is the Prime Mover, but they may differ on the manner and timing. Old-earth creationists affirm that God took His time, while young-earth proponents believe God did His work pronto.

Young-earth creationists are the most vocal, have been thus far the most successful at garnering sympathetic support, and have been hogging the press. What escapes public attention is that there is a quiet group of creationists who are convinced we reside on an ancient planet. From an old-earth creationist point of view, it is the extreme young earthers, the radical right, that are causing untold grief, bringing down scorn on the entire creationist movement.

The Irony of Young-Earth Creationism

C. S. Lewis in his book *Mere Christianity* says the devil "sends errors into the world in pairs of opposites." Lewis concludes:

> He (the devil) relies on your extra dislike of the one error
> to draw you gradually into the opposite one. But do not
> be fooled. We have to keep our eyes on the goal and go
> straight through between both errors. [3]

Nowhere could Lewis be more on target than with this issue. Evolution is being used as a scare tactic, a straw man, to draw gullible Christians into young-earth creationism. The bitter irony about young-earth doctrine is that not only is the science dreadful, even the biblical exegesis is unsound.

According to the young-earth view, the "days of creation" in Genesis 1 are interpreted as 24-hour time periods. From nothing at all to the first humans, Adam and Eve, took less than one calendar week. Claiming a "literal" rendering of the Genesis account, young-earth creationists posit a 6,000 to 10,000 year-old universe with man, in the form of Adam, growing his garden, naming the animals, and having his operation before the paint was dry.

Calling on the "Received Text," Archbishop Ussher used the age of each patriarch at the birth of his son to derive a date for Adam, and the Creation itself six days earlier. Thus, 4004 BC has been used as the starting point for the history of the universe among many young-earth proponents. A few daring rebels have ignored the Archbishop, however, and have pushed the creation date out to a generous 10,000 years, thereby giving a minuscule amount of weight to the evidence from nature.

There are no biological links between species in young-earth doctrine, and no death of any kind occurred in the world prior to Adam's fall. All the animals, and presumably insects, too, would have multiplied and filled the earth, not to mention your vegetable garden, without dying had it not been for Original Sin. [4]

To further impress the biblical mandate upon nature, young-earth creationists call for a vapor canopy, and no rainfall, for the first 1,656 years. Representatives of all the world's animals, including dinosaurs, were loaded on Noah's ark.

Then a world-wide flood arranged the fossil record with complex fossils overlying simple varieties; distributed neat layers of incongruous sedimentary materials; gouged out the Grand Canyon to a depth of two miles; synthesized numerous ice ages; and even moved continents around the globe. And, amazingly, it's all derived from "inerrant" Scripture.

Young-earth advocates disallow evolution primarily because they take away the time for it to happen. As Gish put it:

> It is this author's belief that a sound Biblical exegesis requires the acceptance of the catastrophist-recent creation interpretation of earth history. If this interpretation is accepted, the evolution model, of course, becomes inconceivable. [5]

What escapes those who stand on such declarations is that the reverse is also authorized. If you reject their interpretation of earth history, as virtually every scientist does, then evolution becomes conceivable; and since a "sound Biblical exegesis" is part of the same package, the Bible bears the brunt as the source of those shenanigans.

The technique employed by young-earth creationists is to stand on the Bible (literally, I think), and dredge up whatever supportive inconsistencies they can find in nature. They then assert that the overwhelming majority of the world's scientists are totally in error on earth dating. With their claim that the earth and universe could be less than 10,000 years old, young-earth advocates try to shake the very foundations of our basic sciences.

Although the term "creation science" could have meant advancing scientific evidence of a creation process authored by God, regrettably it does not. The opportunity for that term to have an honorable meaning has been wasted. Creation science is nothing more than the label pinned on young-earth creationists' crude attempts to force fit the formation of the earth and universe into an absurd time frame demanded by their narrow biblical constraints.

It is this attempt at literalism with blinders attached that is a paramount source of conflict. In the young-earth view, the Lord spoke and the world appeared instantly, fully formed, ancient in appearance, and ready for human occupancy.

For a variety of reasons, many Christian fundamentalists have embraced the young-earth position. On the surface it appears to be literal, it's certainly traditional, and it doesn't allow for any of that evolution stuff whatsoever.

Although most of us have at least some familiarity with the science of biology, geology, physics, and math, the "science of creation" probably would draw a few blank stares. What is creation science? The defunct 1981 Arkansas Act used the following definition:

> (a) 'Creation-science' means the scientific evidences for creation and inferences from those scientific evidences. Creation-science includes the scientific evidences and related inferences that indicate (1) Sudden creation of the

universe, energy, and life from nothing; (2) The insufficiency of mutation and natural selection in bringing about development of all living kinds from a single organism; (3) Changes only within fixed limits of originally created kinds of plants and animals; (4) Separate ancestry for man and apes; (5) Explanation of the earth's geology by catastrophism, including the occurrence of a worldwide flood; and (6) A relatively recent inception of the earth and living kinds.

Few creationists would have any trouble with creation science had the above definition been limited to the first sentence. Taken in its entirety, the definition deteriorates into the controversial aspects of young earth and flood geology. Therefore, the term "creation science" excludes all creationists who dare to entertain the possibility that our Creator might have used a longer time to bring all this about.

The contention that the world is young, but only looks old, is totally bereft of earthly or heavenly evidence. Not a shred of credible data from nature has yet been uncovered to support young-earth claims. No radioactive rock has yet been discovered containing so small an amount of decay element that it could be dated to such a young age. Not one fossil bone has been found in Precambrian sedimentary rock. No dinosaur bones have been uncovered in conjunction with human remains, and so forth.

What young-earth proponents have dredged up and presented as supporting evidence has demonstrated either an ignorance of, or a disregard for, the readily available and conclusive evidence that this old planet has been around for billions of years.

Undaunted by the absence of data to support young-earth doctrine, proponents make their points by sowing seeds of doubt. They search for whatever anomalies or inconsistencies they can find in nature in an effort to poke holes in the universally accepted belief in an ancient earth.

The method employed to gain followers is simple. First, young-earth leaders wrap themselves in Scripture. By taking what appears to be a literal Bible interpretation, they gather the support of trusting believers.

Next, young-earth proponents illuminate discrepancies in evolution. We saw in the previous chapter that the theory of evolution has problems of its own, making it an excellent target. Then they trot out paltry evidence, which holds not a drop of water, but can sound credible to the uninformed. Lastly, they travel about, sell books, and solicit donations to further advance their ministry of misinformation.

Creationist Myths

Some of the arguments used to promote young-earth views are as follows:

The Young Moon

In 1960, Hans Petterson wished to calculate the amount of space dust falling on the earth. After collecting samples from a mountain top in Hawaii, he analyzed the nickel content, and deduced that 14 million tons of dust fall on the earth each year. Supposedly, a similar amount was falling on the moon.

According to creationist literature, this figure extrapolated for five billion years would have put a layer of dust on the moon's surface of about 54 feet. [6] That much dust lying on the surface could have made for a hazardous moon landing, and NASA scientists were concerned about that.

When Neil Armstrong and Buzz Aldrin landed on the moon in 1969, they extracted a core sample of about five inches. Young-earth proponents were eager to point to this as "proof" that the moon, and the earth by inference, were young. [7] Subsequent measurements of the moon's surface on the Apollo 17 mission got a seismic reading of the moon's regolith varying from 20 to 120 feet. The regolith is a complicated surface containing dust, ash, and moon rocks. The amount of meteoritic material in the lunar soil varies from about 1.5% to 10%. [8]

Direct reading of outer space dust done by satellites, coupled with Surveyor 3 findings of micrometeorite bombardment impacting on the moon itself, demonstrated that Petterson's initial rudimentary calculations weren't even close. Van Till reports:

> By best current estimates, then, the accumulation of me-
> teoritic dust on the moon would contribute a layer less
> than one centimeter thick in four billion years. [9]

In other words, the amount of dust found on the moon was about right for a 4.5 billion year exposure. As recently as 1987, creationist literature continued to flaunt their long discredited young moon evidence. [10]

The Incredible Shrinking Sun

The contention here is that the sun is shrinking at such a rapid rate that only a scant 20 million years ago the sun's surface would have been touching the earth's surface! [11] This statistic was based on a 1979 report by John Eddy and Aram Boornazian who deduced the shrinkage of the sun's surface at the linear rate of 5 feet per hour from computing the time required for the sun to cross the celestial meridian at noon.

The Eddy-Boornazian data was challenged from the beginning. For one thing, the sun is known to oscillate in size, and this has been borne out from various tests. In 1980, Irwin Shapiro used Mercury transit rates across the sun's surface, and came up with figures about one-seventh the value computed by Eddy and Boornazian.

In 1984, John Eddy teamed up with Claus Frohlich and published another startling revelation. From 1967 to 1980, the sun actually *increased* in size at the linear rate of 8 feet per hour! [12] Thanks to the marvels of science, we now know that 20 million years ago the sun was the size of a mustard seed!

This example is typical of the kind of slapdash science creation "scientists" will use to prove whatever they need to prove that might bolster their case. The shrinking sun argument used in creationist literature continued to be published as recently as 1987 - three years after Eddy co-published data showing exactly the opposite.

The Magnetic Earth

We all know that the earth has a magnetic field around it. When I was an Air Force navigator, we relied on "magnetic north" for compass bearings. The magnetic north pole differs from true north by varying degrees depending on where you are on the earth's surface since they are not located at the same place.

The point young-earth creationists make is that the magnetic field is decaying, getting weaker. This implies that in the past it was stronger. If we could go back in time, we would see an increasingly stronger magnetic field.

According to their statistics, 10,000 years ago the earth's magnetic field would have been "as strong as the magnetic field in a magnetic star." "Surely our earth never had a magnetic field stronger than a star," they say. "That would limit the age of the earth to 10,000 years..." [13]   (Really, I'm not making this stuff up.)

Their supporting evidence comes from young-earth advocate, Dr. Thomas Barnes. In 1973, he took 150 years worth of data showing a gradual decay in the magnetic field. Barnes then plotted an exponential curve to the data points. The Barnes curve showed the magnetic field to be not just decreasing, but decreasing at an increasing rate. He then took his 150 years worth of data and extrapolated it to 20,000 BC! Barnes published the results, but not the data.

Dr. S. G. Brush did publish the data to which he plotted a conventional straight line through the data points. The difference this made is that the same magnetic strength Barnes achieved at 20,000 BC with his curve, took over *100 million years* with a straight line. [14] In essence, Barnes "cooked" the original data to contrive a result compatible with his beliefs.

Miscalculating the rate of magnetic decay was only part of the error. What we find in reality is that the earth's magnetic field has not decayed nonstop for billions of years. It has fluctuated through time in sine wave fashion. Core samples of igneous rocks taken from the earth's surface show that the earth's magnetic field has gone through numerous reversals, swapping magnetic north and south poles periodically throughout the earth's long past. [15]

Please notice in the three previous examples that creation "scientists" will use strictly uniformitarian principles to support their claims. They take a rate, extrapolate that rate into the past, and then deduce that the resultant figure cannot possibly be true, and therefore, the earth must be young.

Now, note what they say about geologists' methods. In *The Genesis Record*, Morris makes a point about the flood ordering the fossil sequence found in sedimentary rocks. He then says:

> Man's perverse and depraved nature has somehow distorted both into a system of evolution and uniformity. 16

"Uniformity" can be defined as a projected continuity. It is the assumption that the rates and processes we see today are the same as in the past. This is simply the most conservative stance you could take. The alternative is to assume that something (who knows what?) caused the rates or processes to change. Since we would not know whether the rates or processes changed up or down, the "no change" assumption is as middle-of-the-road as you can get.

The precautions with taking a uniformitarian approach are that you must have a considerable number of established data points, and that you not try to extrapolate too far. For example, on one day in a particular city the maximum temperature was 72 degrees Fahrenheit, the next day it was 67 degrees. An erroneous uniformitarian assumption would project that a year from now we will be in another ice age, or that last year we were living in an inferno.

The important thing to remember, though, is that what makes uniformity inherently good or bad, from a young-earth creationist's point of view, is directly dependent on who uses it. If they use it - it's good; if evolutionists use it - it's bad!

Where is the Salt?

Another young-earth argument is that not enough salt or minerals are in the bottom of the oceans. Here young-earth creationists base their assumptions on the supposition that if the earth was old, oceans would be 4.6 billion year-old stagnant basins. They ignore the universally-accepted documentation of sea floor spreading and plate tectonics that depict the earth as a vibrant and living surface.

In *Science Held Hostage*, Van Till makes this point:

> In 1954 Goldschmidt provided data and discussion supporting the conclusion that the dissolved material in ocean water is in an equilibrium state, being added and removed at equal rates. [17] *Chemical Oceanography*, published in 1965 and revised in 1975, includes extensive discussions of the processes by which elements are removed from ocean water, although those parts of the work appear to have been disregarded by the young-earth advocates who quote from this book in *Scientific Creationism* and elsewhere. [18]

Misguided Science

The previously cited examples are merely representative of what young-earth creationists try to foist off as "science." The list goes on. In the 1982 McLean vs. Arkansas Board of Education decision, Judge William Overton made this telling comment:

> While anybody is free to approach a scientific inquiry in any fashion they choose, they cannot properly describe the methodology used as scientific, if they start with a conclusion and refuse to change it regardless of the evidence developed during the course of the investigation. [19]

Hugh Ross in his book *The Fingerprint of God* spoke of "scientific creationists" as being "sadly misguided and misguiding many whose science education and biblical training are inadequate to aid them in evaluation." Ross identified four key problems with young-earth arguments. They involve (1) faulty assumptions; (2) faulty data; (3) misapplication of principles, laws, and equations; and (4) ignorance of mitigating evidence. [20]

So, you might ask, why not reach out to those dear ones in brotherly love, and show them gently the abundant evidence for an old earth? If we just present all the data, they will see the errors of their ways and come to enlightenment. Right? Try it!

If you tell them about the speed of light and the billions of light years between earth and distant galaxies, they will tell you the speed of light may have changed. Demonstrate that radiometric dating confirms the ages of rocks to billions of years, and they respond there is no way to know how much decay element God put in when He created the earth in the first place.

Explain that it takes millions of years for caves to form stalactites and massive columns, and they will contend there are stalactites under the Jefferson Memorial in Washington, D.C. Show young-earthers 150 million years of dinosaur history, and you will learn about the angler in South Africa who landed a coelacanth in 1938. [21]

Talk about the layers worn by water erosion in the red wall of the Grand Canyon where those particular layers were once on the earth's surface, and prepare to hear that sub-aqueous water seepage could have caused it.

Point out that there are layers of salt evaporite deposits, hundreds of feet thick, well beneath the earth's surface, covered by sedimentary rock they say was laid down by the flood, and they will respond that underground volcanoes may have boiled the water beneath the earth's surface, evaporating the salt out of the sea water in the midst of a raging, global deluge.

Get the pattern? No matter what evidence, or what bit of data, or what undeniable fact is presented proving the earth is old - and any answer, no matter how incongruous, or how cockamamie, or how illogical, will suffice. And they call it creation *science*?

Old Testament Evidence for an Old Earth

Turning away from general revelation, let us look at special revelation. If an earth of great age is mandated by the evidence from nature, then the inspired Scriptures ought to agree. Rest assured, they do.

In Job 15:1, Eliphaz asked Job, "Wast thou made before the hills?" Does it seem reasonable that Eliphaz would have used this question of digging sarcasm had he thought the age of the hills and the age of man were virtually the same, varying by a scant five days?

The intent of Eliphaz in Job is confirmed by Habakkuk 3:6. The mountains are described as "everlasting," the hills are "perpetual." The Hebrew words *'ad* and *'owlam* mean "long duration," "ancient," "forever," and "continuous existence."

Does the Bible comment on the earth-age dispute? Consider Ecclesiastes 1:10: "Is there any thing whereof it may be said, See, this is new? it hath been already of old time, which was before us." Could "any thing" include an earth, for example?

Sin and Death

Young-earth creationists maintain the Bible prohibits death, even in the animal world, until Adam commits Original Sin. Genesis 3:17b and Romans 8:22 are summoned for oblique support, but essentially this idea of no death in the animal kingdom before Adam hinges on their interpretation of about one half of one verse in Romans.

Citing Romans 5:12, Morris explains that death "entered into the world" only when sin came by man. And, he continues:

> ... it is as obvious as anything could be that the fossil record now found in the sedimentary rocks of the earth's crust could only have been formed sometime *after* man sinned. [22]

Is that what the Bible says? All the world's predator animals had to wait for sin before they could put their claws and fangs into tasty red meat? Can you envision black clouds of hungry buzzards egging Eve on? Does that sound plausible?

Let us get some perspective. What did Paul say in Romans? "Wherefore, as by one man sin entered into the world, and death by sin ..." (Rom. 5:12a). Does this mean Adam's sin caused death in the animal world too? Did sin afflict both man and animals in the eyes of Paul?

There are at least two reasons that death was not dealt to the animal world through Adam's fall. First, the fossil record is replete with over half a billion years worth of animal death. That predates Adam by a wide margin. The second reason is that animals do not belong in the same "world" as man, attested to by the Bible writers themselves.

Notice that Morris did not use the entire verse. He stopped in mid-sentence, in fact. This is what follows the semi-colon. Romans 5:12b: "and so death passed upon all men, for that all have sinned." So the Bible tells us, "as obvious as anything could be," who or what is affected by Adam's sin - men, not animals.

Additional clarification can be found in the following verse. Romans 5:13: "For until the law sin was in the world: but sin is not imputed when there is no law." Who was under the law, men or animals? Did animals tithe, fast, celebrate feasts, honor the sabbath, keep the commandments, or offer up unblemished sacrifices?

In Romans 4:13: "For the promise, that he should be the heir of the world, was not to Abraham, or to his seed, through the law, but through the righteousness of faith." Were animals "through the righteousness of faith" to be joint heirs of the world along with the descendants of Abraham?

Had Adam's sin carried a death sentence for the animal world as well, those concerned critters could take heart with I Corinthians 15:22: "For as in Adam all die, even so in Christ shall all be made alive." So, if young-earth proponents are right, keep stocking those pet cemeteries, there's a new day a-comin'. The trumpet will blow, and millions of years worth of animal life will burst forth!

Putting frivolity aside, Scripture forbids such a ludicrous interpretation of Romans 5:12. Adam's sin of disobedience caused death unto all his generations. In addition, the death referred to in this passage more probably refers to spiritual death than physical death. With continued access to the tree of life, Adam would have lived forever despite Original Sin, according to Genesis 3:22.

Anyone entertaining the slightest, fleeting thought of signing on to young-earth doctrine should weigh heavily the words of a certain apostle: "But there were false prophets also among the people, even as there shall be false teachers among you, who privily shall bring in damnable heresies ..." (II Pet. 2:1).

In II Peter 3:3-8, the apostle warns of "*scoffers*," and affirms the "heavens were of *old*," and "that one day is with the Lord as a *thousand years*."

Ignoring the words of the apostle, young-earth creationists *scoff* at the overwhelming evidence of earth antiquity, insist the earth and heavens are *young*, and assert that one day is with the Lord as *24 hours*!

The Appearance of Age

When they come under a reasoned attack, there is yet another rationalization young-earth advocates will employ. Flinching under the weight of old-earth evidence, they offer an alternate explanation. The world is young, they claim, but was created to look old.

In 1857, Phillip Gosse wrote *Omphalos*, a book advocating that God created false records in nature to date his recent creation artificially. Just as Adam must have been created an adult, Gosse argued, the Creator designed the earth to look old. The earth would be young, but would have the "appearance of age."

The appearance of age argument is similarily flawed, just like the rest of the young-earth argument. Had Adam been aged artificially in the same sense that the universe and earth bear the undeniable stamp of antiquity, a physical exam would have revealed worn dentition, liver spots, scar tissue, callouses, blood cholesterol, wrinkles, and all the other physical signs of an aging adult.

The notion that the universe was brought about with an apparent age, or that it looks old but is really young, crumbles under its own weight. How ironic it would have been for God to have commanded us, "Thou shalt not bear false witness," and have expected us to adhere to a criterion that He would have violated from the very beginning.

From the book of Romans we find we are held accountable by the evidence of nature. "For the invisible things of Him from the creation of the world are clearly seen, being understood by the things that are made, even his eternal power and Godhead; so that they are without excuse" (Rom. 1:20). Had an artificially-dated planet been palmed off on us by a clever sleight-of-hand artist we would not be without excuse, we'd have a great excuse!

Inherent with the appearance of age argument is a classic "Catch 22." [23] If the world is old in complete agreement with the way it looks, then why would God give us a book telling us it is young? And if the world is young, then it had to be manufactured deliberately and cleverly to look old.

Thus, the orthodoxy of young-earth creationism poses an insane dilemma; if the world is old, God would be a fibber, and if young, He would be a counterfeiter! Taking young-earth dogma to its conclusion, if we could not trust God to give us a true history of the world we live in, how could we trust Him to give us true history and true prophecy in His Book? A God who could falsify nature might falsify a resurrection!

Ironically, these implied allegations raised by those who profess to be believers call God's very credibility into question. True words demand true works. "For the word of the Lord is right and all his works are done in truth" (Psa. 33:4).

I can think of only one reason that the earth would look old, and that would be so that when we observe it, we think it is old. If God wants us to think that, then that is what we should think. I would rather believe it looks old because it is old. But either way, we should continue to think that way, since that is what He expects. And if I found any evidence for a young earth, He could trust me not to blab about it. If God had created the earth recently, but wanted to make it look old, the last thing I would do is bring up contrary evidence to point out oversights.

For example, the moon rocks had just the right amount of decay element in the rock samples astronauts brought back to give consistent dates in the billions of years, but, oh no, He must have forgotten to put enough dust on the moon to jibe with the radioactive dates! What must He have been thinking of? All of that contrived young-earth evidence, if true, would call for a bumbling creator!

The perplexing theme of young-earth creationism is not that God made the earth with the appearance of age. Rather, the inferred message is that God made a clever attempt to create an old-appearing earth, but thanks to the efforts of those hard-working researchers, they have uncovered His inconsistencies. So, first they accuse Him of deception, and then they assert He wasn't good at it! What a confused message that is.

A Lost Cause

The theory of evolution and its atheistic bent may have been the initial impetus for young-earth creationism, but once the course was set, no revelations of science have generated even the slightest changes. Young-earth proponents will use whatever argument, spurious or otherwise, if it will advance their cause.

The case for evolution does have a principal requirement. Long periods of time are needed for species to evolve from simple to more complex life forms. If millions of years were available then evolution could have happened. If the time is denied, then gradual evolution would be impossible, and only sudden creation will work.

Defeating evolution at any cost appears to have been the motivation, but now the helmsman has been swept overboard, vested interests have been established, and many are caught up in perpetuating a wayward "ministry."

The tragic enigma is this: when authors who purport to be Bible scholars put forth an erroneous theory they claim is based on "inerrant" Scripture, it is Bible credibility that suffers - Bible error is the conclusion. The lamentable effect is for the baby of biblical truth to be tossed out with the bath water of young-earth creationism.

Christians Caught in the Web

In light of scriptural limitations, coupled with voluminous scientific data which is totally one sided, the question is: Why have so many conservative Christians adopted young-earth creationism? The answer lies in this: while young-earth creationists can be debunked for using flawed logic in this particular area, in other areas of Christian doctrine in general their theology is quite sound. This makes the poisoned pill easier to swallow for eager, well-intentioned evangelicals, hungry for the Word, and angered by the popularity of evolutionism.

The fallacy of young-earth doctrine would be a lot easier to detect if it were not encapsulated in what is otherwise commonly accepted hermeneutics. Ask any major league pitcher, and he will tell you the fast ball sets up the curve. If God's truth is "sharper than any two-edged sword" (Heb. 4:12), then perhaps a falsehood is deadlier than a double-pointed pitchfork. One point wounds evangelicals, impeding their effectiveness as soul winners. The other point pierces the hearts of unbelievers in their rejection of the message delivered by unreliable messengers.

If evangelicals cannot be trusted in a simple matter such as the age of the earth, which can be verified easily, then how could we be trusted on the doctrine of shed blood for the remission of sin, for example, where the corroborative evidence is far less abundant? Therein is the crying shame. The unbeliever remains in unbelief because the Bible is presented as unbelievable right from the first chapter.

Truth for Truth's Sake

Alan Hayward in his book *Creation and Evolution* tells of enduring "10 months of prayerful uncertainty" agonizing over whether he should write a book that would lay bare the fallacies of the young-earth argument. He wrote the book, he said, "because Providence seemed to be urging me in that direction." He concluded:

> Why, then, do I think it necessary to put the case for an
> ancient earth? Partly, I suppose, because I care about truth,
> for its own sake. [24]

In his months of prayerful deliberation, Hayward might have weighed the words of the prophet, Samuel, "Only fear the Lord and serve Him in truth with all your heart: for consider how great things He hath done for you" (I Sam. 12:24).

Truth is the issue. True worshippers worship "in truth" (John 4:23-24). How else could we serve and worship a God of truth, but in truth?

In the days of the Apostle Paul, Judaizers attempted to infiltrate the early gentile churches. Their message was one of subtle deceit. Good, you are followers of Jesus, the argument went. Of course, Jesus was a Jew, and was properly circumcised in accordance with Jewish tradition.

Further, they would argue, Jesus obeyed the laws and customs. He honored the holy days and refrained from eating certain meats. If you wish to be His followers, then you need to take on all the Jewish trappings. If you want to be like Jesus, the Jew, then you need to become as Jews yourselves.

This was a compelling argument to impressionable neophyte believers. Paul, writing from prison, warned the Colossians, "And this I say, lest any man should beguile you with enticing words. Beware lest any man spoil you through philosophy and vain deceit, after the tradition of men, after the rudiments of the world, and not after Christ" (Col. 2:4, 8).

If Paul were with us in the flesh today, he might echo the words in Galatians 3:1 to admonish fundamentalist and conservative Christian churches who have been hoodwinked and deceived by the false teaching of young-earth creationism, "Oh foolish [fundamentalists], who hath bewitched you ...?"

NOTES

[1] Robert H. Wilson, *The Story of Old Town & Gentry Row* (Kennett Square, Pennsylvania: KNA Press, Inc.), 1.

[2] Ibid., 2.

[3] C. S. Lewis, *Mere Christianity* (New York: MacMillan Publishing Company, 1952), 145.

[4] Dick Fischer, "The Bible Proves That Creationism Is Wrong," *The Washington Post* (August 17, 1986), C1.

[5] Duane Gish, *Evolution: The Challenge of the Fossil Record* (El Cajon: Creation-Life Publishers, 1985), 52.

[6] Harold S. Slusher, "Some Astronomical Evidences for a Youthful Solar System," *Creation Research Society Quarterly*, 8 (1971), 55.

[7] Ibid., 55.

[8] Howard Van Till, *Science Held Hostage* (Downers Grove: Intervarsity Press, 1988), 73.

[9] Ibid., 71.

[10] Paul S. Taylor, *Origins Handbook* (1987), 12.

[11] Ibid., 11.

[12] Van Till, *Science Held Hostage*, 51.

[13] Taylor, *Origins Handbook*, 12.

[14] Alan Hayward, *Creation and Evolution: The Facts and Fallacies* (London: Triangle, 1986), 137.

[15] Dan Wonderly, *God's Time-Records in Ancient Sediments* (Flint: Crystal Press, 1977), 222.

[16] Henry Morris, *The Genesis Record* (San Diego: Creation-Life Publishers, 1976), 218.

[17] V. M. Goldschmidt, *Geochemistry* (New York: Oxford University Press, 1954).

[18] Van Till, *Science Held Hostage*, 89.

[19] "McLean v. Arkansas Board of Education," U. S. District Court Judge William R. Overton, January 5, 1982, reprinted in *Science* (February 19, 1982).

[20] Hugh Ross, *The Fingerprint of God* (Orange: Promise Publishing Co., 1989), 155.

[21] Christopher Anderson, editor, "Fishing for Missing Links With a Chain," *Science* Vol. 261 (30 July 1993), 553.

[22] Morris, *The Genesis Record*, 79.

[23] *Catch-22* was a classic book and film about an Army Air Corps navigator named Yossarian who sought to get out of World War II by claiming insanity. The Army doctors explained that since war itself was insane, his wanting to escape it showed he was sane. Those who were truly insane were those who preferred to fight the war, but they stayed in. The only way he could prove himself insane was to stay in the war which thwarted what he was trying to do - namely get out. Thus the "catch 22."

[24] Hayward, *Creation and Evolution: The Facts and Fallacies*, 79.

Chapter 5

THEISTIC EVOLUTION: OKAY, EVOLUTION - BUT GOD DID IT

Many thoughtful men and women have invested their time and talent on this issue of science and Scripture and life's origins. They have rejected the extremes of atheistic evolution at one end of the spectrum and young-earth creationism on the other, and have sought middle-ground solutions.

Currently there are three explanations that have gained followers. All three seek to recognize the sensible parts of the two extremist arguments while attempting to avoid obvious pitfalls. These are Theistic Evolution (sometimes called Continuous Creation or Evolutionary Creation), Gap Restitution, and Progressive Creation.

These centrist theories acknowledge God as the Prime Mover. All recognize an ancient earth, which brings with it evolution as a foregone conclusion, but defined narrowly as change through time. Simple organisms precede more complex life forms in the fossil record, and this infers a continuity of life. A line of descendance can be drawn linking up modern life with archaic ancestors. The degree to which these competing theories ascribe to an "interconnectedness" or a continuity of life through time, and to what degree divine intervention or creation played a part, sets the lines of demarcation between them.

Theistic evolution holds to complete continuity with archaic life forms being ancestral to present-day life, and genetic connections between all surviving animals including man. Gap theory calls for a single break in that continuity, and leaves it open

as to when a termination of pre-existing life and a subsequent renewal took place.

Progressive creation posits multiple gaps occurring in nature
such that God would have been active in creating life forms specially, without ancestors, throughout history. Micro-evolution is
permitted after introduction of a life form, such that some animals
can be related, but progressive creation permits no direct connection between man and any other members of the animal kingdom.

Some degree of change through time has taken place resulting
in life as we know it, and recognizing that is common to all three
centrist theories. The fossil record shows continuity and similarities suggestive of shared common ancestry. The methods of explaining these continuities and similarities define these alternative
explanations.

Recent evidence from genetic research is only confirming what
biologists have long concluded. As a result, shared common ancestry among existing species, perhaps including man, must be
recognized in some way if any proposed middle-ground solution
has a hope of gaining acceptance from the scientific community.
At the same time, evangelicals are not going to endorse a solution
which eliminates the Creator from the creation process, nor should
they.

In one respect we can say that all of life is creation. "Evolution" best describes "change through time," though, and whether
God was active or passive in the process is open to argument. Further, if God was active, then in what way? If passive, what was He
doing? These are questions that may never be answered fully, but
seeking after God is encouraged in the Bible, so it should not do
any disservice to muse upon it a bit.

In a Gallup Poll conducted in November 21-24, 1991, only 9%
of those polled showed any belief that life might have arisen without God's involvement. That same poll showed that slightly more
respondents favored man having ancient ancestors than man having a beginning of only 10,000 years ago, or less. [1]

It is apparent that a middle-ground, centrist position fits what
many believe already about man's origins. An acceptable

solution recognizing God's governance, coupled with an aware-ness of the antiquity of the earth and the antiquity of its inhabit-ants, should be welcomed with open arms. Such a solution might also draw proponents from the extremist warring camps as well.

The difficulty is that these three alternatives all come prepack-aged with disabling liabilities, and fail to achieve a satisfying rec-onciliation. Harmonizing is acutely necessary. All three of these proposed solutions try to do that. Unfortunately, none of them work.

Theistic Evolution

The Jesuit priest and expert paleontologist Teilhard de Chardin described evolution as:

> ... a general condition to which all theories, all hypoth-eses, all systems must bow, and which they must satisfy henceforward if they are to be thinkable and true. Evolu-tion is a light illuminating all facts, a curve that all lines must follow. 2

Teilhard, who died in 1955, saw in evolution a clear direction, a wonderful logic, and looking into the future, envisioned a con-tinuing upward trend. Yet, setting up evolution as "a light illumi-nating all facts" would be regarded by most creationists as a sell-out to what is really just a theory after all.

Teilhard died too soon to see the fulfillment of his prophecy which is just as well. Why shatter a dreamer's dream? Those of us living in the late 1990's have the opportunity to further experience the dubious benefits of continuing evolution.

In 1940, the California Department of Education listed the top problems in their public schools as follows:

1. Talking
2. Chewing gum
3. Making noise

4. Running in the halls
5. Getting out of line
6. Wearing improper clothing
7. Not putting paper in the wastebasket

Fifty years later, in 1990, these "problems" were replaced by another set:

1. Drug abuse
2. Alcohol abuse
3. Pregnancy
4. Suicide
5. Rape
6. Robbery
7. Assault

In addition to the decaying of our culture and values, humans suffer from over 3,000 genetic disorders. If this "upward trend" continues, I think it is problematical whether our society can endure another 50 years of evolutionary perfecting.

The hard reality is that whatever the natural progression of life might have been, the deleterious effects of sin have altered the outcome. As a result, evangelicals have great difficulty with the theory of theistic evolution, even though Christians educated in the physical sciences almost universally accept it.

Biologists, whether they be God-fearing or God-denying, who earn a living studying life, are virtually unanimous in opting for a solution incorporating the interconnectedness of living organisms. Special creation with no biological links between species, whether it's sudden and recent, or intermittent and protracted, runs counter to the accepted pattern of evolution, and so biologists reject it out of hand. Peacocke in *God and the New Biology* said:

> Whatever controversies there may or may not be about mechanisms of evolution and of its speed, biologists are agreed about the fact of evolution itself, that is, that all forms of life, current and extinct, are interconnected through evolutionary relationships. [3]

According to theistic evolution, man and beast, and all of animal life, would be descendant from distant ancestors, and related through common links at branch points. Just as in atheistic evolution, the progression of life appears as branches and twigs of one giant phyletic tree. The only premise in theistic evolution is that God caused it all to take place by His guiding hand.

The biggest problem theistic evolutionists face is what to do with Scripture. Typically, they proclaim the Bible to be "true" as an inspired piece of literature, but the truth stops short of being historically accurate. Instead of Genesis 1 being a chronological sequence of events, the order of presentation becomes in Hyers' words, for example, a "cosmogonic" order, [4] or in Clouser's vernacular a "teleological" order. In an article in *Perspectives on Science and Christian Faith*, Clouser elaborates:

> It shows that the intention of the text was to reveal a *teleological* order to the process of creation, which is not at all the same as either a scientific explanation or a description of what an observer would have seen. [5]

In his "teleological" order the "days" of creation are lined up such that the first day on which God separated the light from the darkness sets up a "precondition" for the fourth day when God created the sun, moon, and stars. God separated the sea from the sky on the second day of creation, which lines up with the fifth day when God created sea creatures and birds.

The third day saw the separation of land from the sea and the creation of plant life, and aligns with the creation of animals and humans on the sixth day. In this teleological order, the first three days of creation set up the conditions of the earth required for life forms ushered in on the following three days.

Hyers uses a similar approach. He divides the account into three "movements" or categories, each containing three elements. He labels these categories *"problem," "preparation,"* and *"population."* He places "darkness" in the problem category, "creation of light (day)" in the preparation category and "creation of sun" in the population category. The "watery abyss" problem is prepared by the "creation of the firmament" and populated by the "creation of birds," and so on. 6

Devices such as these may work to some extent, but is that what Moses had in mind? If Moses, the presumed author, had been such a dunderhead that fish came before water, or cattle before grass, or man before mosquitoes, then a hermeneutical helping hand might be useful. But there is nothing wrong with the order of the creation events when allowances are made for the archaic language and Hebrew syntax. Ross goes further:

> The "higher critics" of the last two centuries have badly misinterpreted the first two chapters of Genesis, and by their error led many astray. Establishing the correct point of view and initial conditions for the Genesis creation chronologies yields a sequence of events in perfect harmony with modern science. This accuracy is too amazing for Moses to have guessed. He must have received divine help. 7

If a "teleological" or "cosmogonic" order had been intended for the first chapter of Genesis we might wonder where else in the Bible did the authors use this technique? Even though the sequence of books compiled in the Bible is not necessarily in the order in which they were written, every book of the Bible, and indeed the rest of Genesis too, presents its information in chronological order, even though we may see some minor variation in the sequence of events in the Gospels. 8

Martin Luther gave clear comment on rearranging Bible verses as a method of accommodation:

> This is our foundation: where the Holy Scripture establishes something that must be believed, there we must not deviate from the words, as they sound, neither from the order as it stands, unless an express article of faith (based on clear Scripture passages) compels us to interpret the words otherwise, or arrange them differently. Else, what would become of the Bible? 9

Why call for some kind of sequence other than chronological for the first chapter of Genesis when it would be out of character for every other Old Testament sequencing of events from Genesis 2 on? Besides, in chapter 9 of this book we will see that a chronological sequence in Genesis 1 works at least as well as anything else.

According to the theistic evolution model, God created life specially at the lowest level over 3 billion years ago, but non-stop evolution operated ever since. This means that all men, including Adam (if there was an Adam), were evolved creatures. Clouser continues:

> ... it (the Bible) is not to be understood as teaching that God made a mud model of a life form with no biological predecessors and blew on it with the result that it came alive, hopped up, and walked around. [10]

Call me ultra-orthodox if you will, but that is exactly what I believe. It may not make perfectly logical sense, but subsequent chapters and Books establish Bible credibility firmly, leaving little room to believe that the Genesis account is so incredible we need to invent devices to get past them. Bringing the non-living to life has precedent in both the Old and New Testaments.

Could we forget that Christ was resurrected from the dead (Matt. 28:6)? Also, Jesus brought forth Lazarus from the dead (John 11:43, 44). Elisha raised a child from the dead (II Kings 4:34, 35), and Peter raised Tabitha (Acts 9:40). They "came alive, hopped up, and walked around" at the time of the crucifixion. "The graves were opened; and many bodies of the saints which slept arose ..." (Matt. 27:52).

If bringing to life what was not alive can be performed not only by Jesus, but by prophets and apostles as well, then it does not sound like something that would be out of the question for the Creator of the universe; and besides, that is what the Bible says. Furthermore, Genesis 2:21-23 details Eve's creation graphically. How could anyone say she came into being through evolutionary processes if they give any credence to the inspired text?

An alternative to an evolved Adam in theistic evolution is simply no Adam. Here the Genesis narrative is relegated to one of three categories: poetry, mythology or allegory. If it takes one of these three approaches to make the first eleven chapters of Genesis palatable, where should we stop? Where does the Bible switch to historical-literal, or does it? If Adam was non-existent, what about Noah, or Abraham, or for that matter, all the named patriarchs? Should we wait until the New Testament before we start attaching full credibility to the Bible?

The difficulty with this approach is that the New Testament authors believed the Old Testament, and quoted from it profusely. Luke traces the ancestry of Christ, naming the patriarchs back to and including Adam. When questioned about divorce, Jesus referred to Genesis 1:27, "Have ye not read, that He which made them at the beginning made them male and female" (Matt. 19:4). Then Jesus quoted Genesis 2:24, "For this cause shall a man leave father and mother and shall cleave to his wife: and the twain shall be one flesh" (Matt. 19:5). (Linking these verses is another reason the "man" created in the first chapter of Genesis is not "mankind," but the Adam of Genesis 2, first of the covenant, Eve's husband and Seth's father.)

A good reason for regarding the early chapters of Genesis as history is because the Bible writers themselves treated it as history. Francis Schaeffer pointed out:

> Both the Old and the New Testaments deliberately root themselves back into the early chapters of Genesis, insisting that they are a record of historical events. [11]

In a number of places in the New Testament, the authors seem convinced that Adam was a real person. Paul stated in Romans 5:14, "Nevertheless death reigned from Adam to Moses, even over them that had not sinned after the similitude of Adam's transgression, who is the figure of him that was to come."

Not only did Paul regard Adam as historical on a par with Moses, but the apostle linked an historical Adam with an historical Christ. From I Corinthians 15:22, Paul draws this parallel, "For as in Adam all die, even so in Christ shall all be made alive."

An historical Jesus mandates an historical Adam. Writing to Timothy, Paul said, "For Adam was first formed, then Eve" (I Tim. 2:13). Besides confirming Adam's historicity, the word "formed," *plasso* in the Greek, leaves no doubt as to how Paul believed Adam came into being.

Using theistic evolution as a bridge between science and the Bible drops us into a quagmire of interpretational difficulties. If we do not take the Bible literally, but believe it exists for some purpose, what is it? Once a literal rendering has been eliminated from consideration, the door stands wide open. Almost anything can walk in. What does the text mean if we are not to trust what it says?

Genesis as Poetry or Mythology?

The pre-existence of Israelite epic poetry has been long recognized. Is it possible that the beginning of Genesis was only intended as poetry, and is not to be taken seriously, notwithstanding literally?

It is true that wondrous creation myths abounded throughout the Near East, and many were in the form of epic poems. These began with the origin of the gods, their genealogy, the beginning of the earth and mankind. They told of friction between the deities, culminating in wars between the gods. [12]

In contrast, Genesis speaks of one God who had no origin, and wrought His creation by His own divine will without conflict or interference from adversarial deities. Cassuto refers to the Genesis account "soaring as on eagles wings" above the creation myths that were in circulation. He continues:

> Not many gods but One God; not theogeny, for a god has no family tree; not wars nor strife nor the clash of wills, but only One Will, which rules over everything, without the slightest let or hindrance; not a deity associated with nature and identified with it wholly or in part, but a God who stands absolutely above nature, and outside of it,

and nature and all its constituent elements, even the
sun and all the other entities, be they never so exalted, are
only His creatures, made according to His will. [13]

It is true that components of Israel's poetic tradition can be
found in Scripture, but this does not mean that Genesis can be dis-
missed as mere poetry. Cassuto reasons:

> Semitic thought avoids general statements. Particularly
> in the case of a book like ours, which was not intended
> for the thinkers and the elect few only, but for the people
> as a whole, including also its common folk, it was proper
> that its ideas should be embodied in the language of con-
> crete description. Hence the Torah made use of the con-
> crete traditions that found expression in the 'Wisdom'
> literature and in the ancient heroic poetry of Israel, and
> drew from them material for its structure. Choosing only
> what it deemed worthy, it refined and purified the se-
> lected matter, and moulded the entire narrative to a pat-
> tern of its own - a pattern befitting its purpose and educa-
> tional aim. [14]

This is the pertinent question: Should the first eleven chapters
of Genesis be included as Scripture in what Paul referred to as
"God-breathed"? If so, we cannot call it poetry simply because we
do not comprehend it. We will see in the following chapters that
Genesis is far too buttressed by science and history to be down-
graded to the status of poetic fiction.

Under the heading "Literary Form," *The Expositor's Bible Commentary* says this about Genesis:

> Except for the scattered poetic sections in the Book of
> Genesis, the overall literary form of the book is histori-
> cal narrative, which is the re-presentation of past events
> for the purpose of instruction. Two dimensions are al-
> ways at work in shaping such narratives: (1) the course
> of the historical event itself and (2) the viewpoint of

the author who recounts the events. This dual aspect of historical narrative means that one must not only look at the event in its historical setting, but one must also look for the purpose and intention of the author in recounting the event. [15]

Young lays it on the line in answering the question, Is Genesis poetry or myth? He concluded:

> To escape from the plain factual statements of Genesis some Evangelicals are saying that the early chapters of Genesis are poetry or myth, by which they mean that they are not to be taken as straightforward accounts, and that the acceptance of such a view removes the difficulties. Some are prepared to say that difficulties about the resurrection of Christ are removed at once if you say that the writers of the Gospels do not mean us to understand that a miracle occurred, and that they are simply giving us a poetic account to show that Christ lives on. To adopt such a view, they say, removes all troubles with modern science. But the truth is that, if you accept such beliefs and methods, you are abandoning the Christian faith. If you act thus with Genesis you are not facing up to the facts, and that is a cowardly thing for Evangelicals to do. Genesis is not poetry. There are poetical accounts of creation in the Bible -- Psalm 104, and certain chapters of Job -- and they differ completely from the first chapters of Genesis. Hebrew poetry had certain characteristics, and they are not found in the first chapter of Genesis. So the claim that Genesis one is poetry is no solution to the question. [16]

Genesis as Allegory?

Wade Seaford acknowledged that there must have been "many populations of people before Adam and Eve," and he concluded:

> Only by allegorical interpretation can theology be
> noncontradictive to archaeology, biochemistry, biology,
> and paleoanthropology. 17

Stepping in to give us an allegorical interpretation Seaford calls
for is Charles Fillmore, who wrote in the forward to his book, *Mysteries of Genesis*:

> The 'Five Books of Moses,' of which Genesis is the
> first, have always been credited to Moses, but that he
> was the author seems doubtful in the face of the many
> stories of creation found in the legends and
> hieroglyphs of ancient Egypt, Chaldea, and other na-
> tions that are almost identical with those of Genesis.
> It would thus seem that Moses edited the legends of
> the ages and compiled them into an allegorical his-
> tory of creation.18

In the rest of Fillmore's book, he interprets Genesis allegorically. One Christian sect has used this book in their school of
Christianity, and has embraced the entire theory of evolution from
atom to Adam without dissent, except to add parenthetically - but
God did it.

Complete acceptance of the theory of evolution with the single
exception of adding in a Creator requires spiritualizing, allegoriz-
ing, or negating Genesis in some way. Meanings beyond meanings have to be fathomed out to explain away what otherwise would
have been an erroneous account. Here Fillmore bails out poor
Moses who, without assistance, would have been just plain wrong.

> The Garden of Eden represents a region of being in
> which are provided all primal ideas for the production of
> the beautiful. As described in Genesis it represents alle-
> gorically the elemental life and intelligence placed at the
> disposal of man, through which he is to evolve a soul and
> body.

The Garden of Eden also represents allegorically the elemental forces named by scientists as composing the invisible, etheric universe that Jesus referred to as the 'kingdom of the heavens' and 'Paradise.' It also comprehends the activity of those forces in man's soul and body that, when quickened and regenerated, make him a master of all creation. 'The kingdom of God is within you.' 'East' represents the within as 'west' represents the without. [19]

Does that clarify the narrative? We might have thought the "Garden of Eden" was a place, and that east and west were compass points. The wise men who proceeded westward to find the newborn babe in Bethlehem were journeying "without" - without direction, perhaps.

The ground out of which Adam was formed "represents formed substance: ideas of Truth of which man is conscious," according to Fillmore. What about the trees of the Garden of Eden?

The 'tree' is the substance that connects mind and body, earth and heaven, represented physically by the nerves. The 'tree that is pleasant to the sight' represents the pleasure derived from ascending and descending currents of life over the nerves. The substance of spiritual thought is the 'food' that is good. The 'tree of life also in the midst of the garden' represents the absolute-life principle established in man's consciousness by Divine Mind, the very center of his being. The roots of the 'tree of life' are centered in the solar-plexus region, and they are symbolized in the physical organism by the nerves of that plexus.

The 'tree of the knowledge of good and evil' represents the sympathetic nervous system whose fruit is sensation. When man controls his feelings and emotions his sensations are harmonized and all his functions are supplied with nerve energy. But when man gives way to the pleasure sensation he consumes or 'eats' of that energy and robs his body of its essential nerve food. Thus excessive sense pleasure and the pain that follows are designated as 'good and evil'. [20]

"Solar-plexus"? "Nerve food"? Perhaps we need not belabor what is painfully obvious, but here is the point. Relegating Scripture to the whims of allegorical interpretation is to give *carte blanche* to anyone who claims to be the interpreter. What are the ground rules for this free-wheeling spiritualizing? Who has the right interpretation? Whose explanation would be suspect?

How can Fillmore be tested on his quasi-intellectual, philosophical rendition? Couldn't anyone come along with an allegorical interpretation of his own and sound just as credible, or incredible? That is the overwhelming drawback in a non-literal, allegorical approach. The truth or untruth of it cannot be tested. In his Genesis commentary, Von Rad said:

> There is no trace of the hymnic element in the language, nor is anything said that needs to be understood symbolically or whose deeper meaning has to be deciphered. [21]

Disavow Genesis?

Another alternative for those who advocate theistic evolution is to write off parts of the Old Testament altogether, but what purpose would that serve? Could we believe that an incomprehensibly intelligent, transcendent Creator would be involved totally in the meticulous design work of His creation, and then care not at all if we knew anything about Him? Or worse, would He let us come into possession of a book about Him which we thought was trustworthy, but which was partially or even entirely untrue?

If our state of awareness about the Creator is to the point where we recognize God's hand in bringing the miracle of life into being, then how could we not also believe that he cares about us to the extent that He wants us to know Him? If the Bible was not the means by which He wants us to know Him, then by what other means? What else do we have?

Summary

Theistic Evolution, then, is not a viable answer for believers. But being aware of a process or operation that allows organisms to change and adapt through time, and recognizing that God used such a process, does take us a step in the right direction.

Being aware of the time involved in getting from brachiopods and trilobites to *Homo sapiens* may indicate our level of science education. And acknowledging that intelligent design needs an intelligent designer may put us on the path to further discovery, but neither of these acclamations will bring us into a harmonious relationship with our Creator.

The Old Testament must be dealt with in some responsible way if we are to comprehend our Designer-Creator God, notwithstanding having a heart that is "perfect toward Him." We simply could not put our full faith and confidence in the integrity of New Testament Scriptures while adhering to a poetical, mythological, or allegorical interpretation of Genesis. We also cannot ignore Genesis. Theistic evolution with its imaginative attempts at reconciliation between Bible and science leaves a legacy of either skepticism or unbelief.

NOTES

[1] Jeffery L. Sheler, and Joannie M. Schrof, "The Creation," *U. S. News & World Report* (December 23, 1991), 59.

[2] Bernard Delfgauw, *Evolution: The Theory Of Teilhard de Chardin* (New York: Harper & Row, Publishers, 1969), 10.

[3] Arthur Peacocke, *God and the New Biology* (San Francisco: Harper & Row, Publishers, 1986), 36.

4 Conrad Hyers, T*he Meaning of Creation: Genesis and Modern Science* (Atlanta: John Knox Press, 1984), 57-71.

5 Roy A. Clouser, "Genesis on the Origin of the Human Race," *Perspectives on Science and Christian Faith*, Volume 43, number 1 (1991), 6.

6 Hyers, *The Meaning of Creation: Genesis and Modern Science*, 69.

7 Hugh Ross, *The Fingerprint of God* (Orange: Promise Publishing Co., 1989), 161.

8 Bruce Waltke, "Historical Grammatical Problems," in *Hermeneutics, Inerrancy and the Bible*, eds. Earl D. Radmacher and Robert Preus (Grand Rapids: Zondervan Publishing House, 1984), 86.

9 *Martin Luthers Werke*, Kirtische Gesamtausgabe (Weimar: Bohlau, 1883), 18: 147.

10 Clouser, "Genesis on the Origin of the Human Race," 7.

11 Francis A. Schaeffer, *Genesis In Time & Space* (Downers Grove: Intervarsity Press, 1972), 15.

12 Umberto Cassuto, *A Commentary on the Book of Genesis* (Jerusalem: The Magnes Press, 1944), 7.

13 Ibid., 8.

14 Ibid., 12.

15 Frank E. Gaebelein, Editor, *The Expositor's Bible Commentary*, Vol. 2 (Grand Rapids: Zondervan Publishing House, 1990), 10.

16 E. J. Young, *In The Beginning* (Edinburgh: Banner of Truth Trust, Publishers, 1976), 18.

17 Wade H. Seaford, "Were There People Before Adam and Eve?" in *The Genesis Debate*, ed. Ronald F. Youngblood (Grand Rapids: Baker Book House, 1990), 163.

18 Charles Fillmore, *Mysteries of Genesis* (Unity School of Christianity, Unity Village, Missouri), 3.

19 Ibid., 34.

20 Ibid., 35-36.

21 Gerard Von Rad, *Genesis: A Commentary* (London: SCM Press, 1972), 47.

Chapter 6

INTERMITTENT CREATION: GOD OF THE GAPS

If the Creator used intermittent creation in the past, a contention held by some, then either one gap or multiple gaps incorporating creation activities could be interspersed by periods of evolution. The one gap method is called Gap Restitution, or Gap Theory.

Gap Theory

In his 1909 Reference Bible, Scofield called for two creations.

> The first creative act refers to the dateless past, and gives scope for all the geologic ages. [1]

Scofield cited Jeremiah 2:23-26 and Isaiah 24:1 and 45:18 to assert that these passages:

> ...clearly indicate that the earth had undergone a cataclysmic change as the result of a divine judgment. The face of the earth bears everywhere the marks of such a catastrophe. [2]

The idea of two creations is grounded in Genesis 1:2, "And the earth was without form, and void; and darkness was upon the face of the deep."

Rendering the word "was" as "became," followed by the Hebrew words *tohu wabohu* translated "without form and void," a destruction or a laying waste of whatever might have been here before could clear the way for a new creation. A primeval cataclysm, linked possibly with the fall of Satan, could have terminated previous forms of life leaving no survivors, just their fossilized bones. The creation account in Genesis would be a re-creation account instead. If Adam was to "replenish the earth" (Gen. 1:28), an extermination might be implied eliminating any embarrassing Pre-Adamites.

Critics of gap theory charge that a massive destruction in the middle of a creation narrative would be a strange order of presentation. Changing "was" to "became" has been judged to be exegetically unsound. In the next verse, when God said, "Let there be light" (Gen. 3:1), would that be taken to mean there was no light for the first creation, or are we to add the word "again" parenthetically at the end of that sentence? Also, to "replenish the earth" is a bad translation of the Hebrew word, *male*. To "fill the earth" translates the phrase more accurately, which leaves the fate of early man an unanswered question.

With only meager help from the Scriptures, gap theory is a shaky attempt to address the problem of what to do with tell-tale fossils. Even though the rudimentary beginnings of gap theory can be traced back to the early church, the modern-day impetus is driven solely by the revelations of the fossil record. Strategically placing the "gap" is the daunting task.

This gap would have to take place between an initial creation and subsequent extermination of archaic life forms, whatever they were, and the creation of fully modern types which start new species, and Adam, of course, who would be the first modern human.

If young-earth creation theory could be likened to sticking one's head in the sand, gap theory is kind of like pulling out one ear. Even if the Scriptures were more accommodating, gap theory fails to resolve a ponderous difficulty.

Where Is the Gap?

One single gap will not begin to explain all the complexities of the creation-evolution issue. The reality gap theorists evade is: where do you put the gap? A total break between all archaic forms of life and what we could call modern life does not exist in the fossil record. Although many present-day species show sudden entry in the progression of life, they do not all enter at the same time.

Positioning this gap at 100,000 years ago to separate modern *Homo sapiens* from any predecessors might be convenient for humankind, but alligators have been unchanged in skeletal form for millions of years. How would alligators have bridged the gap placed strategically to screen out ancestral hominids?

Even placing a gap that would separate modern humans from ancient forerunners is a futile exercise. Archaic *Homo sapiens* go back approximately 300,000 years, and did not die out until about 35,000 years ago. Neanderthals also would have bridged the gap, first appearing in the fossil record some 130,000 years ago, and enduring for 100,000 years before becoming extinct. So even though the biblical support for gap theory is paltry, the fossil record is foreboding.

Different species of hominids overlap in the fossil record. There is a continuity between modern and archaic life forms of all kinds. Millions of years separate the beginnings of the vertebrates: fish, amphibians, reptiles, mammals, primates, and man. These are just a sprinkling of examples that are beyond the explanatory capabilities of one single gap. Gap theory flounders not only because of the lack of scriptural support, or because it calls for some kind of judgment before Adam's fall, but also because even if there were such a gap, no matter where it is placed, it settles nothing.

Alas, gap theory as a means of reconciliation is itself *tohu wabohu* (without form and void). A gap might help explain one problem, though it is hard to know which one, but after playing that gap card, we would still be left with a profusion of unanswered questions. The result is that gap theory has few serious proponents, and has been widely recognized as being of little value as a solution to the complex problem of life's origins.

Progressive Creation

It was pointed out in the previous chapter that all proposed middle-ground solutions recognize the earth's great age and take into account some degree of change through time in animal life and plant life too. Progressive creation tries to de-emphasize the part played by evolution, while placing increased emphasis on the part creation might have played during the past 3 billion years.

In progressive creation, initial life forms such as blue-green algae and bacteria, or their precursors, were results of acts of special creation, something that is held in common with all other creationist theories. And this is a point all creationists can make with confidence. Science has yet to come up with any data that would explain the beginnings of life in a more satisfactory manner.

Shunning complex evolutionary relationships, however, progressive creationists advocate that the procession of life has been sustained throughout the ages by numerous acts of special creation. Any unguided evolution that might have taken place has been limited to micro-evolution only. Progressive creationists reject large scale macro-evolution, which links up all life forms.

Opinions vary as to how God might have performed these creative acts, ranging from out-of-the-dust creations to God having made nucleotide substitutions on the DNA molecules of pre-existing creatures like sliding the beads on an abacus.

However He did it, which is open to dispute, the important point is that, according to progressive creationists, God intervened continually throughout life's history to bring creatures into existence that otherwise would not have lived if only random, unguided, evolutionary processes were in operation. This rationale seeks to explain why there are gaps in the fossil record, and thus progressive creation has been dubbed the "God of the gaps" theory. Evolutionists are quick to point out, however, that gaps are the absence of evidence, not the evidence of absence.

Sudden Appearance of Animal Life

It has long been recognized that various types of animal life have appeared suddenly, and have remained relatively unchanged for up to hundreds of millions of years in some instances. This is called "stasis." [3] Instead of continual, uninterrupted evolution where creatures would have been spinning off new forms of animal life constantly, many kinds of animals seem to stay pretty much the same as they were when they first appeared.

Cockroaches have remained cockroaches, and alligators have stayed alligators, from the time they were first introduced up until today. Progressive creationists are among the first to point this out, and attribute it to a Creator who must have personally ushered in increasingly complex creatures periodically since primordial life first appeared.

Postulating a discontinuous procession of life interspersed with countless acts of special creation throughout history does not sit well with biologists who generally have a distaste for miracles as an answer for anything that can be explained by natural processes. This is true especially when a perfectly plausible explanation has been offered for the gaps in the fossil record, and it can be supported with data.

At least one large scale extinction, the dinosaurs, which disappeared 65 million years ago, has been correlated with a meteor impact. In addition, the earth's climate has swung in pendulum fashion in the past, from ice ages through interglacials, then back to the last ice age about 12,000 years ago, followed by the present interglacial period we live in today.

Animals on islands or isolated continents diverge morphologically from their forbears through time. Some of the extinctions of animal life with subsequent introduction of a new assortment of novel types seem to coincide with "twitches" in the climate. When some of the earth's water becomes locked in ice, the sea level falls, opening up land bridges. This connects previously isolated land masses allowing what had been separated animals to intermix, or one type will overrun and displace another.

Thanks to the new argon-argon isotope dating method, geochronologists are better able to pinpoint changes in the earth's climatological past. As reported by *Science*:

> The short time scale of these events and their clear coincidence with turning points in the evolution of

> mammals are providing the tightest links yet between
> global climate change and evolution on land. 4

In contrast to a rational, natural explanation, the implications of special creation as a method of introducing new life forms is perplexing. If we look at the progression of predecessors that have led to the modern horse, we can see how special creation stacks up against an evolutionary scenario.

A Horse Is a Horse

The first animal designated a horse was originally named *Eohippus* found from the early Eocene about 55 million years ago. Did *Eohippus* pop into life without ancestors? No, it seems he derived from a phenacodont condylarth from which it hardly differs in dentition, though there is a significant difference in brain complexity. 5 *Eohippus* was roughly the size of a lamb with four toes on the front feet and three on the hind feet.

The closely related *Palaeotherium* co-existed with little *Eohippus*, and had more advanced horse-like characteristics, including dentition, like the horses of the Miocene. These palaeotheriids were more like a tapir, though, and could grow as big as a rhinoceros. This example of interrelatedness between coexisting and overlapping life forms is replete in the fossil record in stark contradiction to the theory of progressive creation. 6

The procession of horses from the Eocene up to today is one of subtle, gradual change of feet from three toes to one, in dentition as they progressed from browsers to grazers, and in size as each succeeding type was larger than its predecessor. *Mesohippus* replaces *Eohippus* followed by *Parahippus* and *Merychippus* leading to *Pliohippus*, and finally to modern *Equus*. They grow in dimension. Skull casings and brains become larger. Teeth and first feet and then hooves show steady step-by-step transition. With these animals in particular, no abrupt, morphological leaps exist in the fossil record.

Their feet, which become hooves, are especially interesting. As each succeeding animal replaces its precursor the two side toes move up higher on the leg until by *Merychippus* they stick out at the sides as useless appendages no longer touching the ground. A study of the comparative anatomy from *Mesohippus* to *Pliohippus* shows each animal stands clearly as an intermediate, an unmistakably related link between predecessor and successor.

> Today the genus comprises zebras, onagers, asses, wild and domestic horses. With *Equus* we see the product of 55 million years of adaptation to ever increasing aridity, to more efficient feeding on grasses, to larger overall size and more rapid locomotion. [7]

It's true that the fossil record of the horse also shows intermediate branches or twigs leading to dead ends, but this is consistent with the entire fossil record that could better be described as a phyletic "bush" than a tree.

Hominids

Closer to home, let us look at hominids. We could start with *Homo habilis* 2.5 million years ago, which is totally arbitrary due to the pre-existence of the Australopithicines. Although relegated to ape status, the Australopithicines are far more human-like than is any modern-day member of the genus Pongidae, and paleontologists universally assign them to the hominids, placing them as ancestral to the genus *Homo*. [8]

Homo erectus replaced *Homo habilis* in the fossil progression about 1.6 million years ago, who in turn gave way to *Homo sapiens* when archaic types emerged 300,000 years ago. Again, the picture with humans is much the same as with the horse; subtle changes, advanced anatomy, and the appearance of descent.

Both archaic *Homo sapiens* and Neanderthals make their appearance earlier than what are considered to be truly modern *Homo sapiens*, and they continue to endure for tens of thousands of years after that. [9] This overlapping implies ancestral relationships even though paleontologists may not agree as to who begat whom.

Arguments on the specifics arise among paleoanthropologists. Most prefer straight line graphs of descendance with clear-cut dates and species identifications. Others choose a more complex, bushy pattern, where two or three different species of hominids may have co-existed at the same time.

Whatever hominid descent scenario wins out eventually need not concern us. It is the general pattern of continuity and overlapping leading to modern humans that makes it exceedingly difficult to place an Adam who could be ancestral to us all, and yet have no genetic links to the past.

Genesis Genetics

Progressive creationists equate the word "man" in Scripture with modern man, and assert that Adam must have started the line of *Homo sapiens* sometime in the distant past. The idea is to drive Adam far enough back into prehistory to be believable. Generally they avoid any specific date, since there is no date that can be rationally defended. But for purposes of illustration, Adam could be penciled in about 100,000 years ago to align with the beginning of the modern *Homo sapiens*.

Placing Adam's time frame in the distant past implies the Genesis record omitted the names of hundreds of generations who supposedly lived between Adam and Abraham. The rationale is that the word "begat" does not mean "the immediate father of" in all cases, so the named patriarchs in Genesis 5 and 11 would be only a representative sampling. [10]

Progressive creationists call on the elasticity of Hebrew grammar to enable genealogical stretching. The Hebrew word *ben* for "son" can mean "grandson," "children," or even "descendant." [11] Conversely, "ancestor" can be derived from the Hebrew word *'ab*, which normally means "father." So the means for accommodation are in place, and some Bible scholars have taken this path.

It is a tempting device, but like most temptations should be avoided. Occasional shortcuts are taken in Scripture; Jesus is called "the son of David," for one example (Matt. 1:1). But does this confer license to drive the covenant family, Adam and Eve, back in time? And even if we could, nothing is solved by it.

The urge to force the Bible passages into agreement with what we know about the antiquity of man seems to be as powerful a driving force among progressive creationists as the urge to force the earth and universe into compliance with their interpretation of the Bible is with young-earth creationists, and the result is the same - total rejection by the scientific community.

The Bible Impedes Progressive Creation

No one likes ambiguity. When the consequences are dire - What would outrank eternal salvation in order of importance? - there can be an overwhelming desire to resolve difficulties even by using a ploy.

The ploy in this case is to find inconsistencies in Bible genealogies by comparing Old Testament authors with New Testament authors, and saying, for example: Aha! Matthew dropped three relatives out of Jesus's lineage that are listed in II Kings: Ahaziah, Joash, and Amaziah. [12]

These seeming inconsistencies and allowances in Hebrew grammar somehow establish a precedent making the genealogies in Genesis 5 and 11, and Luke 3 fair game, and therefore, expandable at will. Like many other devices, it will not stand up to scrutiny.

Seth has to be the immediate son of Adam (Gen. 4:25). The identical phraseology which sets Adam's age at the birth of his son, Seth, repeats from Seth to Noah (Gen. 5:3-29). If there are no intermediate generations from Adam to Seth, then that should indicate the same thing down the line.

By comparing the number of years Methuselah lived (Gen. 5:27) with his age at the birth of Lamech (Gen. 5:25), with the age of Lamech at the birth of Noah (Gen. 5:28, 29), and with the age of Noah at the time of the flood (Gen. 7:6), it can be seen that Methuselah died near the year of the flood, presumably before the rain started. That ties in the age of the patriarch at his death with the approximate date of the flood, thereby precluding any additions of time between Methuselah and Noah.

In Jude 1:14, Enoch is "the seventh from Adam," inhibiting additional unnamed patriarchs in the first seven generations. So if there is no space to stick in hundreds of generations from Adam to Enoch, and Enoch's son, Methuselah, died near the time of the flood, that is the *coup de grace* to the expanded genealogies method. Inserting additional time or generations is not a workable proposition from Adam to Noah.

Could the narratives and conversations in the early chapters of Genesis have been handed down by word of mouth *over an eon of time*, and arrive at Moses's door intact, not to mention *inerrant*? If tens of thousands of years, and hundreds of generations stood between Adam and Noah, who knows what relationship Noah's or Shem's version passed down through the Semites would have had with the original?

If the ten generations from Adam to Noah are strictly father and son relationships, as most scholars believe, then three of the patriarchs still alive at the time of Noah also lived during the time of Adam. The description of the relationship between Adam and God, and all of the poignant conversations, were only one step removed through the flood. Thus the narrative was safeguarded.

Practical Considerations

Imagine that we have an auditorium with a thousand men in it, and we pass out a piece of paper to each. A prize is awarded to whoever can name the most forefathers on his father's side. How many could come up with ten names? Of any that could, how many would have omitted his own father or grandfather?

The idea that Noah or Shem would have recorded ten forefathers, detailing the age of each at the birth of their first son, or son of the line of promise, and the age at death, while omitting hundreds of intermediate generations, is beyond reason. There is no justification for postulating intermediate, unnamed generations in Genesis 5. Even if it were theoretically possible to insert extra generations, the specific language used with the age of the father given at the birth of each succeeding son prohibits inserting more *time*. So it is a moot point. Archer maintains:

> ... for even allowing the numerous gaps in the chrono-
> logical tables given in Genesis 5 and Genesis 10 it is alto-
> gether unreasonable to suppose that a hundred times as
> many generations are omitted in these tables as are in-
> cluded in them. 13

Since Scripture quashes putting additional time between Adam and Noah, the only place left to put all that time is between the flood and Abraham. But here progressive creation fares no better.

Due to an insistence that Adam has to be ancestral to every human on earth, the Genesis flood, which does not appear to be global, does have to be "universal" as applied to man. As the explanation goes, the direct descendants of Adam never ventured beyond the Mesopotamian valley, and so every human on earth was obliterated by a local flood except for Noah and his family.

This means that the flood also has to be driven back into prehistory, since all humans regardless of racial diversity must somehow emanate from the eight flood survivors. Therefore, according to progressive creation theory, the flood is also relegated to ancient prehistory. Since Abraham's historical niche is fairly secure at about 2055 BC, all that is required is simply to stuff some 90,000 years in between Noah and Abraham. This presents a host of sticky problems.

Parallel civilizations such as those in Sumer and Egypt have left records which tie into the same period from the flood to Abraham, and encompass only a few hundred years, certainly not tens of thousands. 14

Noah curses Canaan in Genesis 9, and Abraham heads for the land of Canaan in Genesis 12. Could over 90,000 years have intervened between just three chapters? Where is all the history that would have transpired in the tens of thousands of years that supposedly lie between the flood and Abraham?

Asshur is Noah's grandson. He set out to build Nineveh, Rehoboth, Callah, and Resen (Gen. 10:11, 12). Nineveh has been excavated to virgin soil, and shows no trace of civilization before 6000 BC. In fact no city in Mesopotamia shows a hint of civilization prior to 12,000 years ago. That leaves the theory of progressive creation with a lot of time to put somewhere, but no place to put it.

Finally, the background information surrounding Adam and his generations to Noah, and from the flood to Abraham, is far too modern in description to have happened at such an early period in man's history. How would livestock raising and farming (Gen. 4:22) have come before hunting and gathering? Could sophisticated musical instruments (Gen. 4:21) predate simple bone flutes? How could metal working (Gen. 4:22) have preceded the Neolithic (late Stone Age) period? It serves no useful purpose to render the Genesis account incredible to bestow a relevance it already has without these good-intentioned efforts.

Why force something that isn't there? If we believe paleontologists, anatomically modern humans go back some 100,000 years; archaic *Homo sapiens* first appeared about 300,000 years ago; and hominids of some description can be traced back over 2.5 million years with precursors to beyond 4 million years ago. And if we trust the biblical text, Adam fits best at about 5000 BC.

Schroeder addresses this issue in *Genesis and the Big Bang*:

> For the Bible scholar, it is not an easy task to accept as reality that for the past 100,000 years there existed animals such as hominids and that the skeletons of these ancient animals are near replicas of those of modern man. But the fossil evidence is abundant and irrefutable. It is folly, no it is counterproductive, to close one's eyes to this fact. [15]

Summary

Overlapping and continuity of life forms with similar and corresponding characteristics is more suggestive of shared common ancestry than it is of special creation without ancestors. What Bible apologists should try to understand is that evolution fits the evidence of nature better than any creationist format unless we were to say that God works at the level of the DNA molecule. But gradual replacement of archaic life with variant life forms more advanced in design sounds like evolution, and is part of the definition of evolution. So why not call it evolution?

It is unwarranted to postulate that God created life forms, where-upon they flourished for awhile, and then after or just prior to their extinction, He produced advanced types miraculously with no an-cestors, and He did this throughout the entire progression of life on earth. Just like the "appearance of age" covered in chapter 4, this smacks of the appearance of evolution, and so it should be treated with the same degree of skepticism.

Clearly, the data of the fossil record points to evolution (nar-rowly defined), although creation better explains the beginning of the universe and the origin of life. A combination of both creation and evolution works better than either does to the exclusion of the other. But the reality is that none of the old-earth creationist theo-ries that attempt to combine elements of both hold water.

NOTES

[1] C. I. Scofield, *The Scofield Study Bible* (1945), 3.

[2] Ibid., 3.

[3] Norman L. Geisler, and Kirby J. Anderson, *Origin Science* (Grand Rapids: Baker Book House, 1987), 152.

[4] Richard A. Kerr, "When Climate Twitches, Evolution Takes Great Leaps," *Science* (18 September 1992), 1622.

[5] R. J. G. Savage and M. R. Long, *Mammal evolution: an illustrated guide* (New York: Facts On File Publications, 1986), 200.

[6] Ibid., 200.

[7] Ibid., 204.

[8] B. J. Williams, *Evolution and Human Origins: An Introduction to Physical Anthropology* (New York: Harper & Row, Publishers, 1979), 192-208.

[9] John J. Putman, ed., "The Search for Modern Humans," *The National Geographic* (October, 1988), 446-447.

[10] Robert C. Newman and Herman J. Eckelmann, *Genesis One and the Origin of the Earth* (Grand Rapids: Baker Book House, 1977), 111.

[11] Lloyd R. Bailey, *Genesis, Creation, and Creationism* (New York: Paulist Press, 1993), 130.

[12] Paul H. Seely, *Inerrant Wisdom: Science & Inerrancy In Biblical Perspective* (Portland: Evangelical Reform, Inc., 1989), 17.

13 Gleason L. Archer, *A Survey of Old Testament Introduction* (Chicago: Moody Press, 1974), 203.

14 Henry H. Halley, *Halley's Bible Handbook* (Grand Rapids: Zondervan Publishing House, 1965), 85-86.

15 Gerald L. Schroeder, *Genesis and the Big Bang* (New York: Bantam Books, 1990), 175.

Chapter 7

INERRANCY:
TO ERR IS HUMAN

The International Council on Biblical Inerrancy met in Chicago in 1978, and after three days of deliberation, the 250 delegates ratified the Chicago Statement On Biblical Application. In the council's words, the Summit meeting "achieved a major restatement for our time of the historic Christian view of Holy Scripture as canonical revelation from God given in the form of composite human testimony in God's will, works and ways." [1]

The Chicago Statement included a Summary and a set of Articles of Affirmation and Denial. Part of the Preface declared:

> The following Statement affirms this inerrancy of Scripture afresh, making clear our understanding of it and warning against its denial. We are persuaded that to deny it is to set aside the witness of Jesus Christ and of the Holy Spirit and to refuse that submission to the claims of God's own Word that marks true Christian faith. [2]

Following are some highlights of the Summary Statement. God is "Truth and speaks truth only." He inspired Holy Scripture to reveal Himself to us "as Creator and Lord, Redeemer and Judge," and "Holy Scripture is God's witness to Himself."

Although written by men, Scripture is God's own Word, and "is of infallible divine authority in all matters upon which it touches: It is to be believed, as God's instruction, in all that it affirms; obeyed, as God's command, in all that it requires; embraced, as God's pledge, in all that it promises."

The Holy Spirit is the "divine Author" who "authenticates it to us by His inward witness and opens our minds to understand its meaning." Scripture is "without error or fault in its teaching, no less in what it states about God's acts in creation," or "about the events of world history." The statement concluded:

> The authority of Scripture is inescapably impaired if this total divine inerrancy is in any way limited or disregarded, or made relative to a view of truth contrary to the Bible's own; and such lapses bring serious loss to both the individual and the Church. [3]

The Affirmations and Denials section included nineteen articles pertaining to aspects of the inerrancy position. It was affirmed that the Holy Scriptures are the "authoritative Word of God," and "the written Word in its entirety is revelation given by God." That God used human language "as a means of revelation" was affirmed, and it was denied any corruption crept in through human sin.

The Statement affirmed that revelation was "progressive" and denied that later revelation "ever corrects" or "contradicts." Inspiration was said to apply to the entire Scriptures such that no part could have been written by human hand alone, although obviously human hands played a part. While the origin of Scripture was divine, the "mode of inspiration" was declared a mystery.

God utilized the "distinctive personalities and literary style of the writers" without overriding their personalities. Inspiration did not confer omniscience on the writers, but their fallen nature caused no distortion or falsehood. "Inspiration" applies only to the autographic text of Scripture. Copies and translations "are the Word of God to the extent that they faithfully represent the original."

It was denied that any essential element of the Christian faith is affected by the absence of the original autographs, or that this absence renders the asserting of Biblical inerrancy invalid or irrelevant. Scripture was affirmed infallible, inerrant, and "free from all falsehood, fraud, or deceit."

The Chicago Statement denied that "biblical infallibility and inerrancy are limited to spiritual, religious, or redemptive themes, exclusive of assertions in the fields of history and science." It was also denied that "scientific hypotheses about earth history may properly be used to overturn the teaching of Scripture on creation and the flood." Although phenomenal observations of nature, reporting falsehoods, and the rounding of numbers were acknowledged, none of these should impact inerrancy negatively.

The doctrine of inerrancy is "grounded in the teaching of the Bible"; it has been "integral to the Church's faith throughout history," according to the Chicago Statement; and "the Holy Spirit bears witness to the Scriptures, assuring believers of the truthfulness of God's written Word." The text of Scripture is to be interpreted by "grammatico-historical exegesis, taking account of its literary forms and devices," and "Scripture is to interpret Scripture."

Denied was the "legitimacy of any treatment of the text or quest for sources lying behind it that leads to relativizing, dehistoricizing, or discounting its teaching, or rejecting its claims to authorship."

Finally, "a confession of the full authority, infallibility, and inerrancy of Scripture is vital to a sound understanding of the whole of the Christian faith," and "such confession should lead to increasing conformity to the image of Christ." That such a confession was necessary for salvation was denied; however, rejection of inerrancy carried "grave consequences." [4]

Getting Perspective

The Chicago Statement is truly praiseworthy. It establishes clear guidelines for conservative believers to follow. A measure of faith is helpful, that is true, but the doctrine of inerrancy can also be established based upon internal and external proof, and we will see some of that in following chapters. A small point of criticism could be raised in that the framers thought the "image of God" was inherited by all humanity, whereas there are indications in the Bible that it accrues to the redeemed. (More of that in chapter 18.)

One of the charges leveled at the doctrine of inerrancy is that parts of the Bible appear to be contradictory. Although it may look like contradiction, there are times when the Bible comments on a particular subject from different perspectives. In Genesis 7 and 8, the flood narrative starts from God's perspective, then shifts to an ark's eye view by Genesis 8:6, and moves back to the Lord's viewpoint at the end of Genesis 8. This changing perspective can be seen easily.

What may not be so easy to see, and could be a source of confusion, is that some of the principles are presented from both man's perspective and from God's view of things. The old battlegrounds of predestination versus free will, and the Doctrine of the Elect (God chooses who will be saved) as opposed to "whosoever believes," constitute prime examples.

In God's omniscience, the future is known to Him. "For whom He did foreknow, he also did predestinate to be conformed to the image of His Son ..." (Rom. 8:29). The book of life has been written. Those named in it have been elected since before time began. "For many are called, but few are chosen" (Matt. 22:14). From God's perspective, salvation is selective.

At the time of the Puritans, this Doctrine of the Elect led to anxious hand wringing as church members agonized over whether or not their names were recorded in the book of life. They overlooked the point that we cannot gain God's perspective. A beetle cannot envision the world in which it lives the way a soaring eagle can. We humans have only a limited human point of view.

From our perspective, John 3:15 applies: "That whosoever believeth in Him should not perish, but have eternal life." Life everlasting is guaranteed available to everyone. The Book of Life contains all the names of those who believe in God's promise and act on faith. John 11:26: "And whosoever liveth and believeth in me shall never die. Believest thou this?" Everyone is eligible.

Historical Inerrancy

Scripture is the infallible Word of God set down by human authors under the inspiration of the Holy Spirit. It is not a new concept concocted by fundamentalists to counter the recent move away from total inerrancy and toward modern theology, or neo-orthodoxy.

The early church accepted the entire Bible as the inspired word of God. Though not stated explicitly, because it was not really an issue, it was implicit in the positions maintained. There was no reason to believe otherwise. [5] Just the words of Scripture itself were persuasive enough.

In II Timothy 3:16: "All Scripture is given by inspiration of God ..." *Theopneustos*, translated as "inspiration," is the word Paul used meaning literally, "God-'spirated' " or "God breathed-out," wrote Harold Lindsell in *The Battle for the Bible*:

> It means that God indeed is the author of Scripture, and Scripture is the product of His creative breath. The emphasis is not on inspired writers as much as it is on inspired Scripture. Scripture is 'breathed out.' This is not to suggest that the Holy Spirit did not move on the writers themselves, but that the writers produced a product, which, while it was their own, was also the Word of the living God. [6]

The Bible is a book of history, of a people striving and failing in their faltering attempts to follow God. It is a book of miracles, of a Savior born of a virgin, of His excruciating death on a cross, of His resurrection from the grave, and the redemption of mankind through His sacrifice.

Scripture includes do's and don'ts, testifying and teaching, trials and tribulation. The Bible is a book of prophecy, both those that have taken place, and of those still to be fulfilled, including the second coming. And the Scriptures are a source of consolation, revelation, and motivation.

To those who dared to question Jesus's credentials, there were the words of the long deceased prophets who foretold of His coming. For those who doubted His pronouncements, accompanying miracles substantiated His authority. To corroborate miracles, we have the personal testimony of those affected.

It is the synergistic effect of the intertwining of all these elements that culminates in the Bible's profound impact. The history, the prophecies, the miracles, and the doctrine are mutually dependent and mutually supporting, and it all testifies to a living God.

In II Peter 1:21, Peter proclaims, "For the prophecy came not in old time by the will of man: but holy men of God spake as they were moved by the Holy Ghost."

Individual authors were men with different personalities, education, and experiences, and each drew upon his own remembrances; but the work of Scripture was the work of God wrought by human hands, aided by the Holy Spirit, and necessarily free from error. From men who did not possess perfect knowledge came a perfect work. God was the conductor, and His few chosen mortals were His musicians. Paul may have plucked a harp while Peter played a flute, but the melody was the same.

In Paul's letter to the Thessalonians, he affirms the concept of divine revelation, "For this cause also thank we God without ceasing, because, when ye received the word of God which ye heard of us, ye received it not as the word of men, but as it is in truth, the word of God, which effectively worketh in you that believe." (I Thess. 2:13).

In Galatians 1:11, 12, Paul invokes higher authority, "But I certify you, brethren, that the gospel which was preached of me is not after man. For I neither received it of man, neither was I taught it, but by the revelation of Jesus Christ."

The Bible itself proclaims its own authority. It self-evinces its own authenticity and testifies to its own truth. Of course, any book or any author could state flatly that what is written is the truth. But who else would believe it, and who would testify to their claim?

In the case of the Bible, Christ proves through His attitude and declarations that the Scriptures are inspired by God. Hughes gives this insight in *Basic Christian Doctrines*:

And what could be more significant for the Christian than the attitude of Christ Himself (with which, of course, the attitude of the apostles is fully consonant)? He emphasized not only that He had not come to destroy the law and the prophets but to fulfill them, but also that not one jot or tittle would fall away until all things were accomplished (Matt. 5:17 f). The Scripture was for Him something that could not be broken (John 10:35). In the temptation in the wilderness, the devil is on each occasion repulsed, without further argument, by a quotation from the Old Testament, 'It stands written...,' the plain inference being that it is the absolutely authoritative Word of God (Matt. 4:4,7,10).

It was the Old Testament Scriptures, viewed in their entirety--'the law of Moses, and the prophets, and the psalms'--which the risen Saviour expounded to His disciples, emphasizing the necessity that all things written in them concerning Him should be fulfilled (Luke 24:44 ff). Throughout the New Testament, indeed, the whole of Christ's life, death, and resurrection is seen in the light of the fulfillment of Holy Scripture, and therefore as a vindication of the Bible as the inspired Word of God. [7]

In *Studies in Theology*, Boettner affirms:

That Jesus considered the Old Testament fully inspired is abundantly clear. He quoted it as such, and based His teachings upon it. One of His clearest statements is found in John 10:35, where, in controversy with the Jews, His defense takes the form of an appeal to Scripture, and after quoting a statement He adds the significant words, "And the Scripture cannot be broken." [8]

In many ways, from many authors, the Bible is pronounced trustworthy. But if the Bible is God-breathed, God-inspired, and co-authored by the Holy Spirit, can we still derive wrong conclusions? Furthermore, is it possible that error has crept in, polluting and diluting the truth? Sadly, the answer is "yes," and the elementary reason stems from the fact that the Scriptures have been placed in human hands.

Qualified Inerrancy

Explicit in the Chicago Statement were affirmations and denials aimed specifically at critics of biblical authority. It was worded carefully that the human authors working under inspiration wrote down inerrant original manuscripts. Scribes laboring under perspiration were capable of doing what humans persist in doing regardless of safeguards applied. They made mistakes and omissions. They wrote in the margins. Scribal glosses could get caught up in the succeeding text.

> In the Hebrew manuscripts that have been examined, some 80,000 various readings actually occur as to the Hebrew consonants. How many occur as to the vowel-points and accents, no man knows. [9]

The original Hebrew set down by the inspired authors was without vowels or punctuation. Vowel points were added later by the Masoretes. [10] Scrolls wore out. New copies were made from earlier copies late in their life cycle, when they were not as easily deciphered. And so, inerrancy extends to the copies only where they are true to the originals. Although a vigorous defender of the authority of Scripture, nevertheless, Archer adds this disclaimer:

> We are referring here, of course, to the original manuscripts in Hebrew, Aramaic, and Greek; we make no such claim concerning later copies of manuscripts. [11]

Some of the criticisms aimed at the Bible can be accounted for by scribal errors. This is easiest to see by comparing parallel accounts in separate books of the Bible. II Samuel 10:18 credits

David with slaying 40,000 Syrian "horsemen" while the like account in I Chronicles 19:18 calls them "footmen," the latter being more probable. [12]

Had Luke not recorded the name of Cainan (Luke 3:36) we would have no way of knowing he was absent in Genesis 11:12-13. The Septuagint version records Cainan as the son of Arphaxad who lived 130 years before begetting Salah. Cainan is a conspicuous deletion in the Masoretic text confirmed by the New Testament author, Luke.

The mode of inspiration was declared a mystery in the Chicago Statement. One of the controversies among conservative Bible scholars concerns the manner in which Moses compiled the early passages of Genesis. Did Moses use source documents? Did he listen to and set down oral traditions, or did Moses take dictation?

The evidence lies on the side of source documents, inasmuch as Sumerian and Accadian flood accounts are far too similar to the Genesis narrative to escape the likelihood of a source common to all. Since both the Sumerian and Accadian languages precede ancient Hebrew, accounts in those languages must have been recorded earlier. Therefore, it can be assumed that written documents were available to Moses, although conceivably he might have drawn on oral tradition.

The use of *Elohim*, translated as "God" from Genesis 1:1 through Genesis 2:3, contrasted with *Yahweh*, translated as "Lord God" beginning with Genesis 2:4, is just one support for the theory that source documents were used. [13] Writing in 1860, Edward William Lane answered the question: "To whom was the revelation originally communicated?"

> It is held by many (perhaps we might truly say by almost all) of the best Biblical critics, on grounds that appear to us to be such as hardly admit of any other inference, that the book of Genesis mainly consists of a number of distinct pieces or documents, not revealed to Moses, but collected and arranged by him under the guidance of inspiration. [14]

A Matter of Numbers

"Round numbers" were exempted in the Chicago Statement, and the subject warrants discussion as numbers can be a source of confusion. Comparing the parallel accounts in Chronicles against those in the books of Samuel and Kings, there appear to be eighteen such discrepancies. [15] Probably all these can be attributed to scribal mistakes.

Parallel accounts also exist by which we can take measure of the accuracy of the ages recorded for the patriarchs listed in Genesis 5 and 11. The Masoretic text, the Samaritan Pentateuch, and the Septuagint can be compared as follows:

Ages of the Patriarchs
Before the Birth of the Son

Pre-Flood				Post-Flood			
	MT	**SP**	**LXX**		**MT**	**SP**	**LXX**
Adam	130	130	230	Shem (after flood)	2	2	2
Seth	105	105	205	Arphaxad	35	135	135
Enosh	90	90	190	Cainan (in LXX			
Kenan	70	70	170	and Luke 3:36)	---	---	130
Mahalalel	65	65	165	Shelah	30	130	130
Jared	162	62	162	Eber	34	134	134
Enoch	65	65	165	Peleg	30	130	130
Methuselah	187	67	187	Reu	32	132	132
Lamech	182	53	188	Serug	30	130	130
Noah	500	500	500	Nahor	29	79	79
Years to flood	100	100	100	Terah	70	70	70
Totals	1656	1307	2262	**Totals**	292	942	1072

In examining the age of each patriarch at the birth of his son, deleting 100 years (less likely adding 100 years) is the most common discrepancy. This one hundred year disparity in ages occurs with 15 of the 19 listed patriarchs. The three texts all agree only twice, with Noah and Terah.

Is there any way we could know which text is "inerrant," and which two are flawed? Should we let them vote, and ascribe inerrancy to the numbers supported by any two out of the three? In which case, none of the texts is correct. Perhaps the Septuagint version is accurate, and the other two are flawed, or maybe all three have been mauled by copyists.

The point is this: miscopying of numbers in Genesis and other Old Testament books did happen. But so what? This should not impinge on the inerrancy of the original manuscripts. However, it should quench any enthusiasm for assigning historical dates to biblical events such as the flood, or Adam's creation by simplistically adding up questionable numbers. Waltke comments:

> ... we need to take account of the fact that numbers in the Bible are notoriously difficult to accept on face value, especially as given in the received text of Chronicles, and are legitimately the subject of higher criticism. [16]

Rounding out the subject of numbers, the abundant use of "perfect," "convenient" or "prophetic" numbers such as three, seven, and forty are also exempted from the inerrancy standard. This should not detract or cause us to question the legitimacy of whether events actually happened in seven days or seven years, or forty days or forty years, but should alert us to the fact that a supernatural or mystical significance was attached to certain numbers that go beyond the simple numeric values.

Language Problems

Difficulties were acknowledged in the Chicago Statement, including "lack of modern technical precision" and "observational descriptions of nature," but these do not negate inerrancy. An example of one difficulty, phenomenal language, is found in Psalm 19:4-6. The sun is described as rejoicing "as a strong man to run a race," and the circuit is from one end of the heavens to the other.

We modern folk use descriptive language such as "sunset" and "sunrise" too, even though we know a revolving earth creates the illusion. God accommodates Himself to human language. [17] There is one subtle difference, though. It may be phenomenal language to us because we know better, but it was undoubtedly descriptive language to the human authors of Scripture who did not know better. Should that make a difference? The answer is "no," and here is why.

God used human writers without pre-empting their personalities, knowledge, or writing style to produce His Word for them, for us, and for future generations. The message is in the infallible words, irrespective of human ignorance. God did not need to instruct the writers in cosmology or anthropology before they were qualified to pen Holy Writ.

A few present-day Bible expositors have looked for what the writers "meant to say." This infers that we have to look past cumbersome, archaic words, and into the heads of the ancient writers for true meaning. This places the focus in the wrong place. Frankly, we do not care what they thought, or even what they may have written in personal correspondence. As it concerns Scripture, the emphasis should be placed on the inspired words themselves.

The human authors of Scripture used figures of speech enough like ours that descriptions of nature cause very little confusion. They may have thought the sun, moon, and stars revolved around a flat, saucer-shaped earth that rested on pillars. But what they thought need not concern us. What matters is what they wrote. God's Word penned by these authors is elastic enough to extend beyond what they knew to what we know, and beyond us to what will be known by future generations. That is the beauty of the inspired text. It is not bound by the faulty and limited human knowledge that produced it. Daniel makes that clear. Daniel had no idea what he was writing about, and God gave no explanations (Dan. 12:4).

God's purpose for the Old Testament differs somewhat from His purpose for the New Testament, and the languages are suitable for each. Even though the authors were empowered to write beyond what they knew, still they wrote in the style and grammar of their day. [18] The archaic Hebrew was suitable for history and prophecy. The precision of the Greek language was more appropriate for describing the fulfillment in Christ, and to delineate doctrine.

The Hebrews painted word pictures in a language lacking in exactness. Language even affected their manner of thinking. We can see this in Proverbs 6:16, "These six things doth the Lord hate: yea seven are an abomination unto him ..." Then the proverb lists seven. Anyone thinking in Greek or English might respond, "Well, which is it, six or seven?" This seeming contradiction was not a problem to one thinking and writing in Hebrew. Yet, could we imagine Paul making such a statement?

The inexacting Hebrew and Aramaic languages served to veil prophecy in the Old Testament to protect it from the enemy. But when God wanted to spell it out in no uncertain terms, a precision language was employed, New Testament Greek. Reading it all in English, we may forget this distinction, and expect the Old Testament to adhere to the same standards of comprehensibility as the New Testament. It will not work. The languages are different, the purpose is different, and the ground rules are different.

Furthermore, a few of the words have been altered by time. Some words have connotations different today than when they were first written. The two words that have most contributed to continued misinterpretation of Genesis 1-11 are the Hebrew words *'adam* for "man," and *'erets*, meaning either "land" or "earth."

There are words in our vocabulary today that were absent in Hebrew for obvious reasons; i.e. quasar, planet, asteroid, photon, microbe, cancer, oxygen, estrogen, bytes, carburetor, and hundreds more. And some words are unchanged in definition, for example, dust, eye, bread, tree, air, rib, wife, blood, etc. Definitions for the words "man" and "earth" are caught in between. They are both changed and unchanged. What makes these words tricky is that they mean *somewhat* the same thing today, but are not identical in connotation to what they meant over 3,000 years ago.

The meaning of the word "man" has been broadened in recent years to include more than our unique species. Now "man" also can encompass early forerunners who walked upright, such as *Homo habilis*, *Homo erectus*, archaic *Homo sapiens*, and Neanderthals, none of which can be foisted off on the ape family. The same is true for the word "earth." Even "sun," "moon," and "stars" can be defined with greater complexity today than could have been done in ancient days.

If we could journey back in time, and tap Moses on the shoulder while he was pondering the Red Sea, we might ask, "Pardon me, Moses, I know this is an awkward time for a trivia question, but did you mean to include Neanderthals as 'men,' and did you use the word 'earth' as one of the nine planets in our solar system?"

Assuming we could get Moses's attention, what answer should we expect? Possibly, "What's a Neanderthal? What's a planet? And where did you get those funny looking pants with Levi's written on them?"

In other words, we are on our own. We have to weigh the matter, and decide for ourselves what is intended. Physicists still cannot define the word "gravity" beyond describing it as an attractive force that accompanies matter, coupled with a mathematical equation to measure its strength. We do not know yet how to describe the phenomenon beyond that. If we ever do understand gravity better, the word will not change, only the definition.

Concerning the word "man," Paleontologists refer to man broadly to include all hominids back to, but excluding, the Australopithicines. Archaeologists speak of "modern man" beginning with farming techniques at around 10,000 years ago. At the other extreme, some have tried to define "man" so narrowly as to include a God-consciousness, thereby ruling out not only apes and all early precursors to modern man, but by such a restricted definition, they try to revoke "manhood" for any present-day atheists as well.

In order to decrease confusion, the term "biblical man" could be used beginning at the time of Adam, and would include those living in the same vicinity. This would differentiate them from "non-biblical men" to include all those who were living outside Adam's immediate environment at the time, as well as those long perished.

Thus, Neanderthals are "men," in the broad sense, but not in the biblical sense. The Australian Aborigines (to choose a distant example for illustrative purposes) are "men" by any definition, but "non-biblical," at least in Old Testament times.

All human populations who were distant from the Mesopotamian basin at the time of Adam, Noah, and then Abraham, are in the "non-biblical" category. The circle of inclusion widens gradually after the flood as Bible personalities and prophets encounter more remote populations.

After the creation narrative, the looking glass of the Bible starts with the Garden of Eden at Genesis 2:8. The scope widens to encompass Mesopotamia by the flood, then widens further to the Near East beginning at Genesis 10. Not until the New Testament does the entire populated world come into focus. Christ tells his disciples to preach the gospel "to every creature" (Mark 16:15), thereby welcoming all the world's human populations.

Truth and Error

One of the acute problems is that words have been processed on the journey from the scrolls penned by the inspired authors to the minds of present-day readers. A long chain of scribes and translators has intervened since the authors set down the inspired text, until the English-speaking reader picks up the Bible today, available in different versions.

Our understanding of inspired Scripture can be impeded by three primary sources of error: transcription, translation, and interpretation.

Transcription

When inerrant Scripture is asserted, it includes the caveat "in the autographs." The words set down by human authors originally were inspired by God and inerrant. But the originals have long turned to dust, and copies are all we have left. This is good. If we had the original manuscripts, they themselves could become objects of worship, and likely would become religious relics like the much-traveled bones of Peter.

Hebrew scribes reproduced copies of the sacred text with great care. It was their duty to protect their work conscientiously. They employed safeguards to insure the quality of the finished product. Yet despite precautions, mistakes occurred, since no two copies are alike.

The foreword to the Revised Standard Version (RSV) justifies itself by pointing to discrepancies in the King James Version. In the preface to the RSV, an "incalculable debt" is acknowledged to the early version, but, it continues:

> Yet the King James Version has grave defects. By the middle of the nineteenth century, the development of Biblical studies and the discovery of many manuscripts more ancient than those upon which the King James Version was based, made it manifest that these defects are so many and so serious as to call for revision of the English translation. [19]

The preface to the RSV refers to "errors in copying," "revisions," and how the KJV "suffered in transmission." Nothing on earth remains eternal, however, and the RSV has long been superseded. It is tempting to thrust tongue in cheek, and ask which version is the "true" version, but we should resist impertinence. Suffice it to say that all the versions have merit, and comparing them can aid our overall understanding.

Gesenius observed the original text of the Old Testament has "suffered to a much greater extent than former scholars were inclined to admit." He categorized transcription errors as either unintentional or intentional:

The causes of *unintentional* corruption in the great majority of cases are:

> --Interchange of similar letters, which has sometimes taken place in the early 'Phoenician' writing; transposition or

omissions of single letters, words, or even whole sentences, which are then often added in the margin and thence brought back into the text in the wrong place; such omission is generally due to homoioteleuton, i.e. erroneous repetition of letters, words, and even sentences; its opposite, haplography; and lastly wrong division of words, since at a certain period in the transmission of the text the words were not separated. --*Intentional* changes are due to corrections for the sake of decency or of dogma, and to the insertion of glosses, some of them very early. [20]

Translation

The next step where error crops up is in the translation from the original languages into English. The Hebrew word *har*, for example, can mean "mountain," "hill" or "hill country." In Genesis 7:19, 20, the same word is in both verses. Noah's flood covered the "hills" in the first verse, whereas the "mountains" were covered in the second verse. Also in Genesis 8:5, the word "hills" could have been used just as easily as "mountains."

In Deuteronomy 1:7, *har* is used twice, and again it is translated both ways; first as the "mount" of the Amorites, but as "hills" in the middle of the phrase, "in the plain, in the hills and in the vale." The Hebrew word is again translated "hills" in the phrase, "land of hills and valleys" (Deut. 11:11).

In Joshua 9:1 and 10:40, the "hills" and valleys, and "hills" and vale, could have been rendered "mountains" due to the ambiguity of the Hebrew word, but it was not. *Har* appears three times in I Kings, and has been translated as "hills" all three times. In Isaiah, *har* is "hills" both times. In Psalms 68, 80, 95, 97, 98, 104, and 121, *har* appears nine times, and is rendered "hills" every time. Why was this Hebrew word rendered "mountains" in Genesis 7:20 and 8:5? For the same reason that "earth" was chosen instead of "land," and "heaven" preferred over "sky."

The Bible translators, in the service of King James in the early 1600's, already *knew* the Genesis flood was global at the outset. They fell into the trap of allowing their foregone conclusions about the "global deluge" to drive the translation. The translators were not guided solely by Hebrew verbiage and syntax, but also by presumptive bias.

Psalm 49:1-2 is a case in point: "Hear this all ye people; give ear all ye inhabitants of the world: both low and high, rich and poor, together." The two Hebrew words *bene 'adam*, translated "low" in the second verse, are literally - "sons of Adam"! What comes to us as "high" is the Hebrew *bene 'ish*. *'Ish* is a more general term meaning "man," "male," "human being," or "mankind." Instead of "low and high," which bear no semblance of meaning from the original Hebrew, either "sons of Adam and sons of man," or "Adamite and Non-Adamite" would have been literal translations, faithful to the Hebrew text. Ah, but who could the sons of man be who are not sons of Adam? Non-Adamites? How could that be possible? So, the translators of the authorized version avoided certain controversy by substituting the benign "low and high," virtual synonyms for "poor" and "rich." Modern translators of newer versions have simply followed along.

This technique of substituting words of convenience where *'adam* and *'ish* are contained in the same sentence is used also in Psalm 62:9, where we do not read, "Surely vanity are the sons of Adam, a lie are the sons of man ..." Instead we read, "Surely men of low degree are vanity, and men of high degree are a lie ..." In Isaiah 2:9, do we see the Adamite bow down, and the Non-Adamite humble himself? No, we see instead, "And the mean man (*'adam*) boweth down, and the great man (*'ish*) humbleth himself ..." This same pattern is repeated in Isaiah 31:8, where the term for generic man *'ish* becomes a "mighty man," while *'adam* is a "mean man." So, thanks to a sanitizing translation process, even a prophet may have difficulty getting his message to the people.

Had the Bible translators of the early 1600's known more about global geography, geology, linguistics, paleontology, or anthropology they might have selected their English equivalents with greater precision. If they had, we would not be struggling today to extricate ourselves out of the hermeneutical hole they dug for us.

Interpretation

The first rule of interpretation comes from Exodus 3:5, "Put off thy shoes from off thy feet, for the place whereon thou standest is holy ground." We should be heedful that God's Word is "holy ground." Those who have been called to handle it should do so in fear of the Lord. Though we are all free to interpret for ourselves, when any of us takes a step toward interpreting for others, it is a solemn obligation that carries with it an extra measure of accountability.

Unless there is a good reason that obviates some other approach, we should use a literal interpretation lest we color the text unnecessarily. In the simplest terms "literal" refers to the ordinary meanings of words and phrases without unnecessary embellishment or spiritualization. Of course, metaphor, hyperbole, analogy, symbolism, and anthropomorphism have their place in the Bible, and are not intended to be taken literally.

Regardless of the pot holes that have befallen the inspired text in transcription and translation over thousands of years, it is in the area of interpretation where carelessness can exact a terrible toll. Human rationality rears its sometimes ugly head. It may be that sinful man just cannot grapple with the Word of God without mangling it.

It may be difficult sometimes, but we are obliged to be faithful to the text. When we encounter an unreconcilable difficulty with a passage of Scripture, we need to hold that text in abeyance, and give the Bible the benefit of the doubt based upon the sheer weight of its total credibility.

For the Mathematically Minded

The following equation may be useful in seeing what must take place for original truth, inspired by God, written through the prophets, to yield similar truth in the heads and hearts of today's hearers.

By way of equation: The percentage of truth in the original autographs, minus the percentage of errant transcription, minus the percentage of errant translation, minus the percentage of errant interpretation, equals our confidence level in the biblical truth expressed as a percentage. Displayed in column form:

> Percentage of inspired truth in the autographs
>   minus the percentage of errant transcription
>   minus the percentage of errant translation
>   minus the percentage of errant interpretation
> Individual confidence level expressed as a percentage

Using the formula, let us start with the 100% God-inspired truth of the autographs. If the amount of transcribed text contains just 1% error, for example, and the translation into English was also achieved with only 1% error, and we could comprehend accurately all but 5% of the text, we would have an overall confidence level of 93%.

> 100% God-inspired truth
> - 1% errant transcription
> - 1% errant translation
> - 5% errant interpretation
> 93% overall confidence level

In other words, errors are cumulative. Hopefully, the 7% of resultant error will be in non-critical areas. But in reality, I think the figures for the illustration are low. The combined transcription and translation errors exceed 2% probably, and it is doubtful there are many who comprehend at a 95% level.

This is part of the reason we need to temper our interpretation of Scripture with God's general revelation. What does nature have to say? Weighing in relevant data from science and history can add balance and keep us from going awry.

It is helpful to keep in mind also that understanding varies from individual to individual. Just getting a message intact from sender to receiver can be difficult. Here is an illustration under the heading, "I'm sure that what you think you heard is what you thought I said, but I'm not so sure that what you thought I said is what I think I meant."

In the mid-1960's, a Strategic Air Command B-52 bomber was flying a routine training mission. An instructor pilot was on board, and took over the controls when the pilot, also called the Aircraft Commander, went back to the latrine. The copilot proceeded to rock the plane back and forth to add a degree of difficulty. The aggravated pilot grabbed the intercom and yelled, "Cut it out!"

The navigator and radar navigator had been wondering why the plane was rocking, and to them, the command sounded like "get out!" So they did, activating their ejection seats. This action alarmed the rest of the crew, and they followed suit. When the Aircraft Commander emerged, he was in command of an empty aircraft, much to his chagrin, and so he joined the rest of the crew in vacating a perfectly good B-52.

After the investigation, an abrupt change was made to the emergency procedures section in the check lists of all SAC air crews. The command in the event of an actual emergency was changed to: "Bail out! Bail out! Bail out!" How many planes were spared a similar demise will never be known.

The point is: we need not "bail out" of perfectly good Scripture either. The first eleven chapters of Genesis do not accord with our world view upon a casual reading of the text. We will see, however, that this seeming discrepancy is more apparent than real.

Conservative - Liberal Dichotomy

Although heresies and schisms dogged the church through the ages, it was not until 1859 with the publication of Darwin's book that a significant move toward existentialism began. Emanating from Barth and Bultmann and their followers, a change was instituted away from a truth rooted in the infallible Word of God to a nonpropositional, nonverbal view of inspiration.

Berkouwer believed the Bible may not always be correct, but such inconsequential errors were not detrimental. He also maintained that although God cannot deceive, nevertheless, He can make "mistakes"! 21

Neo-orthodoxy concentrates more on a personal relationship with Christ who reveals any necessary truth in all matters of import directly to the heart of the believer. [22] Therefore, the need for unerring truth is abolished. To liberal proponents, cutting the bonds from the "errant" Old Testament in this manner affords freedom for the New Testament.

The present-day push toward liberalism has been driven partially by difficulties in resolving the words of Scripture with the revelations of science and history, with the Bible taking the hits. This has tended to segment theologians into one of two warring camps: those who mistrust the Bible reside in one camp; those who question science are in the other.

We have at least six choices available when we find history or science in apparent conflict with what we have accepted as a correct interpretation of Scripture. We can do one of the following:

1. "Stonewall it," and cling to our time-honored Bible interpretations, ignoring all contradictory evidence.

2. Compartmentalize, and place the truth of the Word and the truth of the world into sealed mental compartments, forbidding any spillover that could damage our faith in either our religious compartment or in our world view compartment.

3. Look for solutions. Search for areas of agreement with the expectation that ultimately all truth should be reconcilable.

4. Ignore the dilemma completely due to lack of knowledge, or curiosity, or imagination.

5. Go with the flow. Accept academic wisdom at face value, and modify our stance on biblical accuracy accordingly.

6. Relinquish the faith altogether, and regard the Bible as too incredible to be believed.

Conservatives tend to lean toward the first method, while the fifth method has been attractive for liberals. Some prefer the second option, choosing a kind of religion-science schizophrenia as a better alternative to abandoning the faith or looking silly in front of sophisticated family members, colleagues, and friends. For myself, the third alternative has been fruitful, while many churchgoers use the fourth method. Far too many today are making the last choice, a tragic result of the Bible-science controversy.

Leaps of Faith

Conservatives have established their priorities firmly. Since God must come first in the lives of believers, surely His Word does too. Infallible Scripture is the hard foundation rock upon which conservatives stand, daring any competitors to take a counter position. The burden of proof falls on the pretenders to establish whatever credibility they think they can muster. A conservative has an unassailable position constructed soundly on the inerrant Word of God.

Scientific theories may gain acceptance as confirming data supports them, but usually one or more odd bits of information fall outside the parameters, preventing total confirmation. Rarely does science prove anything. The Bible is a well-tested and time-honored product. In light of this, conservatives to their credit have maintained a high view of Scripture despite the persistent onslaughts of higher criticism.

More could have been done, however, to modify counter-productive, dogmatic interpretations that are in disharmony with history and science. This non-compliant attitude has weakened efforts to reach well-educated unbelievers. Continual pontifications on the authority, inerrancy, and infallibility of the Bible largely are being ignored by non-believers and believers alike. Conservatives, in general, have been reluctant to come down from their ivory towers, venture into the blood and guts arenas, and come to grips with stark reality.

Posturing, orating, and proclaiming the inspired Word of God as true has its place, but skirting the tough issues has drawn this zealous band of scholars into a shouting match they are losing due to their steadily dwindling numbers. From Pinnock:

> Conservative Christians have a definitive tendency to minimize the force of Biblical difficulties, just as liberal Christians tend to exaggerate them. Often the orthodox stalwarts simply do not seriously confront well-formulated critical issues. This cannot be excused. These difficulties cannot be swept aside in a flood of rhetoric. [23]

Liberals, on the other hand, have kept their heads out of the clouds that have obscured the conservative's view of things. Everyday, evidence continues to mount that demands attention by serious theologians who wish to avoid dinosaur status. Had many liberals not been convinced of the truth of the Bible they would have abandoned it altogether. It is to their credit that they have stuck to the incorruptible doctrines of Scripture, while seeking areas of accommodation and workable methodologies.

On the negative side of the ledger, liberals for the most part have sold out to science pronouncing the Bible as trustworthy in religious matters, but relieved of any responsibility when it "blunders" into areas of science and history. For the most part they have given up on the concepts of infallibility and inerrancy, and have adopted the critical-historical method in a conciliatory move to re-establish relevance.

Acting prematurely, they have relinquished too much too soon. Declaring the Bible true in religious matters, but superseded by science and history when the Bible strays into these areas is a mixed message that can leave followers in an uneasy state. Furthermore, this accommodating technique opens the door to apostasy where even what the Bible states on religious matters becomes suspect.

Common Ground

One of the purposes of this book is to establish common ground where liberals and conservatives can embrace in harmony. (So call me a Pollyanna.) Yet, both sides must cast off the shackles of presumption that have thus far been a hindrance to progress, and look with new eyes at solving an old problem.

Thank you, conservatives, for your unflagging commitment to the integrity of Scripture. And thank you, liberals, for your willingness to explore new avenues for bringing a relevant gospel to hungry believers. Both sides are owed a debt of gratitude.

Now, let us examine the common ground together. In the remaining chapters, we will explore a solution. By this point, the need for it should be conspicuous.

NOTES

[1] From a flyer put out by the ICBI introducing the Chicago Statement and announcing the 1986 Summit meeting.

[2] Ronald Youngblood, ed., "The Chicago Statement On Biblical Inerrancy," *Evangelicals and Inerrancy* (Nashville: Thomas Nelson, 1984), 236.

[3] Ibid., 231.

[4] Ibid., 231-235.

[5] Robert D. Preus, "The View of the Bible Held by the Church: The Early Church Through Luther," *Inerrancy*, ed. Norman Geisler (Grand Rapids: Zondervan Publishing House, 1980), 357.

[6] Harold Lindsell, *The Battle for the Bible* (Grand Rapids: Zondervan Publishing House, 1976), 34.

[7] Philip E. Hughes, Chap. 3, "The Knowledge of God: The Inspiration of the Bible," *Basic Christian Doctrines*, ed. Carl F. H. Henry (Grand Rapids: Baker Book House, 1962), 16.

[8] Loraine Boettner, *Studies in Theology* (U.S.: The Presbyterian and Reformed Publishing Company, 1947), 18.

[9] Rev. Prof. Moses Stuart, *Critical History and Defense of the Old Testament Canon* (Andover, 1835), 192.

[10] Bruce Waltke, "Historical Grammatical Problems," *Hermeneutics, Inerrancy and the Bible* (Grand Rapids: Zondervan Publishing House, 1984), 79.

[11] Gleason L. Archer, "Alleged Errors and Discrepancies in the Original Manuscripts of the Bible," *Inerrancy*, 59-60.

[12] Ibid., 60.

[13] Charles F. Kraft, *Genesis: Beginnings of the Biblical Drama* (U. S.: Women's Division of Christian Service, Board of Missions, The Methodist Church, 1964), 24-25.

[14] Edward William Lane, Reginald Stuart Poole, ed., *Genesis of the Earth and of Man* (London: Williams and Norgate, 1860), 2-3.

[15] Archer, "Alleged Errors and Discrepancies in the Original Manuscripts of the Bible," 61.

[16] Waltke, "Historical Grammatical Problems," 86.

[17] John H. Gerstner, "The View of the Bible Held by the Early Church: Calvin and the Westminster Divines," *Inerrancy*, 388.

[18] Waltke, "Historical Grammatical Problems," 72.

[19] RSV, (London: Thomas Nelson, 1952), v-x.

[20] Gesenius' *Hebrew Grammar* (London: Oxford University Press, 1910), 21-22.

[21] Gerstner, "The View of the Bible Held by the Early Church: Calvin and the Westminster Divines," 409.

[22] J. I. Packer, "The Adequacy of Human Language," *Inerrancy*, 204.

[23] Clark H. Pinnock, "Baptists And Biblical Authority," *Evangelicals and Inerrancy*, 155.

Chapter 8

THE DAYS OF CREATION: HOURS OR EONS?

A sticking point in reconciling science and the Bible concerns the "days" of creation in Genesis 1. Are God's creative days to be taken as long periods of time, eons, or short periods of time, to wit - 24 hours? We will examine the biblical evidence for applying a long term definition to that stumbling block of a word, "day" in English, *yom* in Hebrew, and see that in the Bible, as in life, it is all a matter of timing.

After God divided the light from the darkness, "God called the light Day, and the darkness He called Night. And the evening and the morning were the first day" (Gen. 1:5).

Considering Moses was the likely author, compiler, or editor of the Pentateuch, was it his intention to convey a day's period of time in this and succeeding passages in exactly the same measure as a man's day? Or was a day of God's time intended, which could contrast as sharply from our measure, as man in the flesh contrasts against God Himself?

The stages of God's creation are revealed in sequence. The first day began when the sun ignited and the first dazzling light struck the primitive planet Earth. On the second day, the Lord divided the waters; vapor or mist was in the air, and liquid covered the surface. Dry land and vegetation came about on the third day.

The sun, moon, and stars were designated time keepers on the fourth day. Day five was devoted to creating the world's fish and fowl (sometimes translated flying creatures). Land animals came on the scene, and man made his appearance on day six. The Lord rested on the seventh day.

And now for the 64 thousand dollar question. How many ticks of the clock took place? If we had the opportunity to view a video tape replay of the entire creation sequence, how much time should we budget? If we took a week's vacation could we fit it all in? How much time are we talking about?

Defining "Day"

The Hebrew word *yom* has the same meaning as "day" in English. It can mean the daylight portion of a day, the entire 24-hour period, a time of undesignated length, or a day of celebration. Which usage did Moses intend in the first passages of Genesis? Better yet, what meaning did God intend to convey through Moses?

Many have come to believe that interpreting those creative days as long periods is a relatively modern phenomenon dictated by the recent findings of science, i.e. sedimentation rates, radioactive decay rates, a vast and expanding universe, and so forth. But such is not the case.

Some of the early church fathers took their cues from Scripture alone without the benefit of a wealth of mitigating evidence from nature. Irenaeus, Origen, Basil, Augustine, and Thomas Aquinas, to name a few, argued that the days of creation must have been long periods of time solely from their understanding of the biblical text. [1] There are some today, however, who advocate that the creation days in Genesis were strictly 24 hours in duration.

Henry Morris in *The Genesis Record* states, "... the Biblical record itself makes it plain that the days of creation are literal days, not long indefinite ages." And further:

> If he (the writer of Genesis) wished to convey the idea of
> long geological ages, however, he could surely have done
> it far more clearly and effectively in other words than in
> those which he selected.

This is the rationale used by Morris:

As though in anticipation of future misunderstanding, God carefully defined His terms! The very first time He used the word 'day' (Hebrew *yom*), He defined it as the 'light,' to distinguish it from the 'darkness' called 'night.'

> Having separated the day and night, God had completed His first day's work. 'The evening and the morning were the first day.' This same formula is used at the conclusion of each of the six days; so it is obvious that the duration of each of the days, including the first, was the same. Furthermore, the 'day' was the 'light' time, when God did His work; the darkness was the 'night' time when God did no work--nothing new took place between the 'evening' and 'morning' of each day. 2

The Ryrie Study Bible follows the same line:

> And there was evening and there was morning, one day. Better, 'day one.' Later Jewish reckoning began the day with eventide (Lev. 23:32). This may be the reason for the order here, or it may simply mean that one day-night cycle was completed. Since daytime closes at evening and the night ends with the morning, the phrase indicates that the first day and night had been completed. Evening and morning cannot be construed to mean an age, but only a day; everywhere in the Pentateuch the word 'day,' when used (as here) with a numerical adjective, means a solar day (now calibrated as 24 hours). 3

Not every theologian or Bible scholar believes humans were walking around on terra firma a scant 144 hours after the Big Bang, or whatever spectacular event which resulted in a molten ball that became our earthly habitat. But the keys to interpretation are not found by scrutinizing Scripture with the world's logic which can be faulty, or with its knowledge which is incomplete, but by comparing Scripture with Scripture itself.

Presuming Moses was the human instrument, and he used *yom* for a creative day, what was he talking about? For the answer we need look no further than to the Bible itself. Oh sure, we could sneak a peek at nature. Like reading a mystery novel, we can skip to the last pages and discover the butler did it, and then read the book knowing from the beginning who the culprit would be in the end. And the abundance of geological and astrophysical evidence underscoring only one answer - an old earth - is a heavy persuader. But the Bible can be gauged on its own terms. Indeed, what better measure?

Can One Day Equal Six Days?

Following the six days of creation and God's sanctification of the seventh day of rest, a shift of focus begins at Genesis 2:4: "These are the generations of the heavens and of the earth when they were created, in the day that the Lord God made the earth and the heavens."

Here Moses used the word "day" as a coverall to apply to the previous six days of creation. But how can one 24-hour day equal six 24-hour days? This is not a problem in semantics, this is a math problem.

If a day of creation is a time of indefinite length, then one large time of indefinite length could equal six smaller times of indefinite length. What happens when we slice a pie into six pieces? The word "pie" could apply to the whole, or to each piece. But one 24-hour day cannot equal six 24-hour days.

To use a "24-hour period" inappropriately as a definition for the word "day" when that word has a variety of meanings, puts Scripture at odds with Scripture when it is totally unnecessary. Attempts to be literal with some passages, while ignoring other passages, may make the Bible appear to be contradictory when that is not the case at all. Archer concludes:

> ... it is abundantly clear that "yom" in Genesis 2:4 cannot possibly be meant as a twenty-four hour day--unless perchance the Scripture contradicts itself! [4]

As a matter of interest, the operative Hebrew word in this passage is not the word "day" (*yom*), but rather "generation" (*toledah*). According to Hebrew lexicons, the word *toledah* always pertains to a long time period, never to such a short span as a mere week. And since the word is plural, we know with certainty that "generations" must refer to multiple periods of time, each of which are longer than a calendar week. [5]

Days of Varying Length

If we take Genesis 2:4 literally, the entire creation event from day one through six is defined by the author of Genesis as a sequence of long periods of time, not a sequence of 24-hour days. Furthermore, those time periods need not be equal in length.

In everyday English usage, just as in Hebrew, the word "day" is used frequently for varying amounts of time. Here is a quick illustration:

A wife greets her husband at the door as he comes home from work, eager for the supper she has prepared. "How was your day?" she asks.

"It was awful," he complains, "I had to work the whole day, I didn't even have time for lunch."

Now is there anyone who would think this man labored nonstop for 24 hours? Of course not, he is talking about a work day. From the context, anyone would know that his "day" lasted probably eight or nine hours. It is the context surrounding the word that determines meaning, not the word taken in isolation.

William Wilson's *Old Testament Word Studies* sums up the possible variations:

> A day; it is frequently put for time in general, or for a long time; a whole period under consideration ... Day is also put for a particular season or time when any extraordinary event happens ... [6]

The "days" of creation certainly do appear to be periods of extraordinary happenings which fit "a long time" definition better than a 24-hour definition.

What about Ryrie's argument that the word "day" in conjunction with a numerical adjective means a solar day? The theologian J. Oliver Buswell answered that question as he replied to another author trying to use a similar line of reasoning:

> It may be true that this is the only case in which the word day is used figuratively when preceded by any numeral, but the reason is that this is the only case in Scripture in which any indefinitely long periods of time are enumerated. The words 'aion' in Greek and 'olam' in Hebrew are literal words for 'age,' but we do not happen to have any case in which God has said 'first age,' 'second age,' 'third age,' etc. The attempt to make a grammatical rule to the effect that the numeral preceding the word day makes it literal, breaks down on the simple fact that this is the only case in all the Scriptures, and in all Hebrew language, I think, in which ages are enumerated one after the other. There is no such rule in anybody's Hebrew grammar anywhere. The author of this objection, or the one from whom he has attempted to quote, has simply put forth with a sound of authority a grammatical rule which does not exist. [7]

Days Without Sun?

Another trouble spot concerns the luminaries.

"And God said, Let there be lights in the firmament of the heaven to divide the day from the night; and let them be for signs, and for seasons, and for days and years:

And God made two great lights; the greater light to rule the day, and the lesser light to rule the night; He made the stars also.

And the evening and the morning were the fourth day" (Gen. 1:14, 16, 19).

On the first day God created light, yet the sun, moon, and stars were not visible until the fourth day. This is no problem to a creation taking roughly 12 billion years to unfold. It would be some 7-10 billion years after the inception, commonly known as the hot big bang, or simply the Big Bang, which brought not only light, but heat and noise as well, before the sun would form and switch on eventually to become our energy and light source. Prior to that, the earth was "formless and void," and darkness prevailed according to Genesis 1:2.

One might think that in the young-earth version, six 24-hour days punctuated by intervals of daylight and darkness would be hard to come by, since they claim the sun was not *created* until the fourth day. This is no deterrence if your mind is made up. Quoting Morris:

> The formula may be rendered literally: 'And there was evening, then morning --day one,' and so on. It is clear that, beginning with the first day and continuing thereafter, there was established a cyclical succession of days and nights--periods of light and periods of darkness.
>
> Such a cyclical light-dark arrangement clearly means that the earth was now rotating on its axis and that there was a source of light on one side of the earth corresponding to the sun, even though the sun was not yet made (Genesis 1:16). It is equally clear that the length of such days could only have been that of a normal solar day. [8]

"Clear" ? He's got to be kidding. And what does Morris mean by a "source of light on one side of the earth corresponding to the sun," but which *wasn't* the sun? Are we to believe that the Creator set up a giant, cosmic spotlight or laser beam to light up the earth for 72 hours before He energized the sun? Doesn't this cast the Creator and His creation in a somewhat artificial light?

Here is another example of perfectly credible Scripture being made to appear incredible through faulty interpretation sprinkled with literalism. If the first four days of creation are periods of time of indefinite length, as many theologians maintain, and not 24-hour periods, as some would have us believe, then the sequence of events becomes far more plausible.

When the Lord created the heavens and the earth, the earth condensed into a fiery, molten ball. Water was vaporized as steam surrounding the superheated globe. Although the sun, moon, and stars were in place and functional, dense clouds would have obscured their view.

We have no way of knowing when the sun ignited, but certainly the sun's energy was required to facilitate photosynthesis for the vegetation that began on the third day. Finally, the earth cooled to where the water vapor in the atmosphere condensed, whereupon the sun, moon, and stars shined through.

An alternate explanation is that sighted creatures began to use the luminaries to measure time on the fourth day. An earth-bound observer who could have witnessed sunset and sunrise did not exist through the first four days of creation. Archer comments:

> Genesis 1:14-19 reveals that in the fourth creative stage God parted the cloud cover enough for direct sunlight to fall on the earth and for accurate observation of the movements of the sun, moon, and stars to take place. Verse 16 should not be understood as indicating the creation of the heavenly bodies for the first time on the fourth creative day; rather it informs us that the sun, moon, and stars created on Day One as the source of light had been placed in their appointed places by God with a view to their eventually functioning as indicators of time ('signs, seasons, days, years') to terrestrial observers. The Hebrew verb 'wayya'as' in v. 16 should better be rendered 'Now [God] had made the two great luminaries, etc.,' rather than as simple past tense, [God] made. [9]

Instead of the word "create" in the passage cited by Archer, a different verb was used meaning "made" or "had made." This makes good sense. The Lord created heaven and earth on day one, but on day four the celestial bodies were available for earthly observers to use as measures of time.

Not only is the word "day" defined by usage in Scripture, the words "evening" and "morning" are also resolved. In Psalm 90, humans are likened to grass. "In the *morning* it flourisheth, and groweth up; in the *evening* it is cut down, and withereth" (Psa. 90:6). Perry Phillips comments:

> I know of no grass that literally springs up in the morning and then is dead by the same evening. Rather, the psalmist has in mind the life cycle of grass in the Levant, which begins its growth with the November rains and dies with the hot, dry, March, desert winds. In this psalm, therefore, "morning" stands for the period of growth and "evening" stands for the period of death. [10]

Fiat Days?

Some interpreters treat the creation days as 24 hours in duration and in proper sequence, but not necessarily consecutive. As this theory goes, God commanded, "Let there be light," but the implementation following His command could have taken any amount of time. Proponents argue that the "evening" and "morning" applied only to His divine fiat or command, not to the entire events which followed.

According to this view, the first 24-hour day may have occurred 4 1/2 billion years ago (12 billion years ago if we start at the Big Bang). The second 24-hour day was, oh, a billion years after the earth was established, with the third 24-hour day taking place two and a half billion years after that, and so on. Separating those six 24-hour periods with at first billions and later millions of years is simply a harmonistic device which seems to distort the data. First of all, it does not take 24 hours to say, "Let there be light." Also, just as the sun never set on the British Empire in the days when "Britannia ruled the waves" and its colonies ringed the globe, likewise the sun never sets on God.

Transcending Time and Space

Sunset and sunrise are visual phenomena. For one to see or experience evening or morning requires that such an observer be in a fixed position on one of the planets which revolve around our sun, in this case Earth.

God is not fixed in a compartment of time or space. We do not view the world as He views it. If any of us could be observers standing on the bright side of the Moon, a la Neil Armstrong, we could see the small hand on our watch make numerous revolutions without witnessing a sunset or a sunrise.

Morning (*boquer*) may have an absolute meaning on God's scale, yet have only symbolic meaning on our scale. Conversely, an early morning sunrise may be an absolute experience to us, yet be symbolic to God, as His sunrise is perpetual, just as His sunset is perpetual, likewise His day is perpetual, as is His night. Thus a 24-hour day definition is not just erroneous, it is superfluous.

For us terrestrials, we cannot envision the world as God must see it. Not only is God omniscient, He is omnipresent. God has the unique characteristic or capability of being everywhere at once - at an infinite number of locations simultaneously. We are situated both in time and space.

When we speak of evening and morning, we consider it synonymous with sunset and sunrise. This is because, as observers, we are locked in time and space at one specific geographical location at any given time. Astronauts, when they are in orbit around the earth, are less restricted. They see many "evenings" and "mornings" during a 24-hour day as they watch the sun disappear and reappear rapidly over the horizon.

From God's perspective, the earth is always half in daylight and half in darkness perennially, day after day from the moment He created it until the moment of its ultimate destruction. In one respect, we might say that God never sees "evening" or "morning," or maybe we could say that He sees an infinite number of "evenings" and "mornings" every single day. But we cannot trap God in Jerusalem or Jericho, and think that He like us is limited to one sunrise and one sunset per 24-hour period, lest we get trapped like Henry Morris and think God cannot work at night.

To human observers, who did not exist until day six, the term "morning" could be thought of as sunrise. The same thing may be true with the lower "living creatures," on day five. But from day one to day four, God's timing alone applied, unmitigated by any human or animal observers. We have no reason to reach any conclusions as to what timetable might have been operating for at least the first four days.

Satan's Fall

The fall of Satan, and with him one third of the heavenly hosts (Rev. 12:3, 4), had to occur before the creation of Adam, as Satan was lurking in the garden ready to make his move the first chance he got.

The whole story of Satan's fall has to be pieced together. From Ezekiel 28:14-15, he was the "anointed cherub," was "on the holy mountain of God," and was "perfect in his ways" from the day he was created until "iniquity" was found in him. His heart was "lifted up" because of his "beauty." From Revelation 12:7-9, we learn there was "war in heaven." Satan fought against "Michael and his angels," and was "cast out into the earth."

How long a period of time lay between the creation of this "angel of light" until his pride overcame him, and he was cast down to earth? Barring two creations, we would have to shoehorn the entire saga into less than 144 hours (6 days x 24 hours), if we were to ascribe to the days of creation being 24-hour periods.

The Sixth Day

At the beginning of day six, the Lord creates "cattle and creeping things and beasts of the earth" (Gen. 1:24). Adam and Eve were created before God rested on the Sabbath Day.

Had that sixth day of Creation been a 24-hour day, it would have been absolutely jam packed with activity. First, God made land animals. That is the easy part. In Genesis 2:20, Adam has to name them all, that is: all "cattle, all the fowl of the air and every beast of the field." If that sounds like a large task for one day, consider this: In the young-earth creationist's model, death does not occur in the world, even among the animals of the world, until Adam commits Original Sin. [11] The species of animals that exist today comprise less than 1% of what has inhabited this planet since it began. Over 99% of all the species which once roamed the earth are now extinct.

If you can get a grip on the magnitude of what a naming problem would be like with the thousands of species in existence today, multiply those species one hundred fold, and then lay those hoards on Adam just a few moments after his first drawn breaths.

Why, it would have been a sight to behold! Turtles and giant sloths would have galloped by like cheetahs and gazelles. Adam would have been chanting out names like a Tennessee auctioneer. Keep in mind, he also would have been cultivating the garden in his spare moments (Gen. 2:15).

No wonder he was looking for a helper (Gen. 2:20), but it would be easy to see why none was found in that first day's blur of activity. And as a perfect ending to a busy day, Adam has a portion of his side removed, and a wife presented to him (Gen. 2:21-22). That certainly would have kept his first few hours on earth, well, interesting. Bewildered and exhausted, Adam would have spent the first night wondering what the next day would bring. Archer comments:

> It must have required some years, or, at the very least, a considerable number of months for him to complete this comprehensive inventory of all the birds, beasts, and insects that populated the Garden of Eden.

> Finally, after this assignment with all its absorbing interest had been completed, Adam felt a renewed sense of emptiness. Genesis 2:20 ends with the words 'but for

Adam no suitable helper was found.' After this long and unsatisfying experience as a lonely bachelor, God saw that Adam was emotionally prepared for a wife--'a suitable helper.' God, therefore, subjected him to a deep sleep, removed from his body the bone that was closest to his heart, and from that physical core of man fashioned the first woman. Finally God presented woman to Adam in all her fresh, unspoiled beauty, and Adam was ecstatic with joy.

Archer concludes:

> ... it has become very apparent that Genesis 1 was never intended to teach that the sixth creative day, when Adam and Eve were both created, lasted a mere twenty-four hours. In view of the long interval of time between these two, it would seem to border on sheer irrationality to insist that all of Adam's experiences in Genesis 2:15-22 could have been crowded into the last hour or two of a literal twenty-four-hour day. [12]

Adam, in apparent agreement with Archer's conclusion, exclaims in Genesis 2:23, "This is now bone of my bones, and flesh of my flesh ..." The word "now" in this passage is the Hebrew *happa'am* usually translated "now at length" or "at last." This term would be appropriate after a long wait or a lengthy search, but not if Eve had been presented to him only a few hours after he was created.

Day of Rest

Even if a 24-hour period could be construed for any one of the first six days of creation, it would not work for the seventh. Here again, Scripture would have to contradict Scripture just to fit an unjustified presupposition. The New Testament refers to the Lord in His rest continuing from the end of creation on through both the Old and New Testaments.

In Hebrews 4:3, "For we which have believed do enter into rest, as He said, As I have sworn in my wrath, if they shall enter into my rest: although the works were finished from the foundation of the world." According to Archer:

> ...that seventh day, that 'Sabbath rest,' in a very definite
> sense has continued on right into the church age. [13]

If the seventh day, the Lord's day of rest, is a long period of time encompassing thousands of years as implied by Scripture, then consistency demands that the first six days be given similar treatment - that is, ages or eons, but not 24-hour periods.

A passage in Exodus has been used as a proof text to bolster the 24-hour day definition, but here again the Bible is uncooperative. "Remember the sabbath day, to keep it holy. Six days shalt thou labor, and do all thy work: But the seventh day is the sabbath of the Lord thy God: in it thou, nor thy son, nor thy daughter, thy manservant, nor thy maidservant, nor thy cattle, nor thy stranger that is within thy gates: For in six days the Lord made heaven and earth, the sea, and all that in them is, and rested the seventh day: wherefore the Lord blessed the sabbath day and hallowed it" (Exod. 2:8-11). Buswell gives the answer:

> If we had no other example of Moses' language, this pas-
> sage might be taken as evidence for a twenty-four hour
> creative day, but we have Scriptural evidence that Moses
> made a radical distinction between God's attitude toward
> time and the attitude of man. What Moses is saying, in
> the total Scriptural context, must be understood as teach-
> ing that man should observe a periodicity in the ratio of
> work to rest, of six days to one day, because God in the
> creation set an example of an analogous periodicity of
> six and one of his kind of days. Surely the fourth com-
> mandment gives no right to say that God's days always
> must be understood to be of the same length as man's
> days, when we have so much evidence to the contrary. [14]

Just as God labored for six days and rested on the seventh, so should man rest on the seventh after six days of work. That is a lesson drawn from analogy. Even as God rests, so should man rest. That does not mean that God and man are on the same time-table.

Days of a Thousand Years

In Psalm 90:4, Moses said, "For a thousand years in Thy sight are like yesterday when it passes by, or as a watch in the night [three to four hours]."

These words leave no doubt that God's timing and man's timing are not to be confused, nor will any simple equation rectify the discrepancy. We have neither the information nor the brain power to figure out what His time might be in relation to our time.

In case we missed the point in Psalms, we have another chance in II Peter. After the apostle declares that false prophets and false teachers will come in the last days, he warns in II Peter 3:5, "For this they willingly are ignorant of, that by the word of God the heavens were of old ..."

Who says the earth and heavens are young? Those who are "willingly ignorant." And to drive the point home, the apostle follows in II Peter 3:8, "But, beloved, be not ignorant of this one thing, that one day is with the Lord as a thousand years and a thousand years as one day." Can it be plainer than that?

Clearly, man's measurements are puny yardsticks indeed. How long is a day of God's creation? We are not told how long it is, we are told how long it is not! Specifically, His time and our time are dissimilar. A 24-hour day is the one interpretation eliminated as a possibility. In the words of Augustine, they were "God-divided days," not "sun-divided days."

Summary

To reiterate:

1. The Hebrew word *yom* has a number of meanings which allow a time of long duration to be an appropriate application of the word "day" without any stretching of the credibility of Scripture.

2. If the sun's appearance is not until the fourth day, it could not have been used as a means of measuring the length of that one or the previous three days.

3. The sixth day of creation is too loaded with events to be stuffed into 24 hours.

4. The seventh day continues on into the church age.

5. We are informed expressly, both in Old and New Testament, that God's time is not to be confused with man's time.

Any thoughtful person who would examine the scriptural evidence alone should be able to conclude that a day of God's creation was not intended to be interpreted as a 24-hour period. But when the scientific evidence is weighed in, making the case convincingly for an ancient earth, then all argument should end.

The days of creation were the periods of time God took to accomplish His creative acts. Interpreting those days as 24-hour time periods violates God's Word in Scripture, and His work in nature, not to mention man's good common sense.

NOTES

[1] Hugh Ross, *The Fingerprint of God* (Orange: Promise Publishing Co., 1989), 141.

[2] Henry M. Morris, *The Genesis Record* (San Diego: Creation-Life Publishers, 1976), 54-55.

[3] Charles C. Ryrie, *The Ryrie Study Bible* (Chicago: Moody Press, 1976), 7.

[4] Gleason L. Archer, *Encyclopedia of Bible Difficulties* (Grand Rapids: The Zondervan Corporation, 1982), 63.

[5] Ross, *The Fingerprint of God*, 151.

[6] William Wilson, *Old Testament Word Studies* (McLean: Macdonald Publishing Co., 1978), 109.

7 Pattle P. T. Pun, *Evolution: Nature and Scripture in Conflict?* (Grand Rapids, Zondervan Publishing House, 1982), 269.

8 Morris, *The Genesis Record*, 55.

9 Archer, *Encyclopedia of Bible Difficulties*, 61.

10 Perry G. Phillips, "Are the Days of Genesis Longer than 24 Hours? The Bible Says 'Yes!'" IBRI *Research Report* No. 40 (1991), 3.

11 Morris, *The Genesis Record*, 79.

12 Archer, *Encyclopedia of Bible Difficulties*, 68.

13 Archer, *Encyclopedia of Bible Difficulties*, 62.

14 J. Oliver Buswell, Jr., *A Systemic Theology of the Christian Religion* (Grand Rapids: Zondervan, 1962), 1:144-45.

Chapter 9

GENESIS ONE AND THE BIG BANG: TWO SOURCES, ONE STORY

It was reported that the Scottish geologist, Hugh Miller, was driven insane in his attempts to reconcile the geologic column with the Genesis narrative. Hopefully, in this chapter we will fare a little better.

Probably all of us have heard or read that the creation narrative in the first chapter of Genesis conflicts with the modern Big Bang theory of cosmology. If scientists are correct about the origins of the universe and the earth, and if the Bible is true history, then there should be no disagreement.

If it can be demonstrated that the biblical text is not contradictory to science, then the reverse should also be true; the Big Bang should present no challenge to the Bible. We will see the value of a literal interpretation of Genesis from the remarkable harmony between Moses's account and the latest theories of modern science. Allegorical, poetical, or mythological pandering is not only unwarranted, but detrimental to the spirit of the text.

If we keep in mind that the six days of creation are long periods of indefinite length where God did His creative work, and if we allow for the limitations of archaic language, then the two accounts, one from Scripture and one from nature, are entirely compatible.

Day One

Genesis 1:1: "In the beginning God created" the sky and the land.

What? That doesn't sound familiar? All right, if you insist, "the heaven and the earth."

Satisfied? Yes, it sounds better in the King James Version, but here is the point. We modern folk are quite educated about our physical universe. At the mention of the word "heaven" we can picture or imagine planets, comets, black holes, supernovas, dark matter, galaxies, and the like. The Hebrew shepherd of 3,000 years ago saw a bright sun in the daytime and twinkling pinpoints of light, accompanied by a moon at night. His "heaven," in the natural sense, was merely our "sky."

This globe is decorated with continents, oceans, islands, and ice caps, but these would be as foreign to that Hebrew shepherd for the word he knew as "earth" as it would be if we tried to foist our known universe on him as "heaven." The "earth" was all the land he could see from horizon to horizon, plus whatever distant lands he may have heard about, nothing more.

"Heaven" and "earth" are correct words, there is nothing wrong with them, but the very first sentence of the Bible can set us up for error if we are not cautious. We need to pause and reflect on a context quite ancient. With that caveat, let us look closely at the beginning.

"In the beginning was the Word, and the Word was with God, and the Word was God" (John 1:1). A word spoken by the Creator to commence the creation might have sounded something like "kaboom!" What had never been before, now was. Space, time, matter, and energy began. God spoke, and a universe came forth.

Astrophysicists say that hydrogen and helium gas spread out in an enormous cloud that through time was destined to become galaxies, stars, a sun, an earth, and us. What has bothered some physicists is that an inception point, a beginning, indicates something they may not wish to admit. The Big Bang can best be described as a creation event, requiring a Creator.

Over the course of billions of years after the Big Bang, clumps of matter drawn together by the force of gravity formed stars revolving around central cores that astronomers believe are black holes. Our own Milky Way is a spiral galaxy with our sun situated in a quiet spot near one of the outer arms. Our solar system, comprised of the Sun and its nine planets, revolves around the center of the Milky Way, making one circuit every 230 million years. [1]

The Beginning of the Universe

If our universe had a beginning, a creation, and its commencement was in an explosion, then we might expect there would be some remnant evidence attesting to that. And there is. Over a dozen independent proofs are available today confirming a beginning point for our universe. [2] We will discuss just six.

Red Shift. When light emitting objects are moving away from us, the light waves are elongated. The visible light is then "shifted" toward the red end of the spectrum. The reverse would also be true, such that if objects in space were speeding toward us, the light waves would be "blue shifted" due to a compacting of the light waves. The greater the speed, the greater the shift. The amount of red shifting affords an approximate measure of the speed at which space objects are propelled apart.

By measuring red shift, Edwin Hubble observed that the farthest galaxies were moving away from us at the fastest speeds. Working backwards, the implications were that the entire universe originated at one place at one time. Thus far, no one has come up with any acceptable alternative, though a few keep trying.

Thermal Heat Loss. The extreme temperatures achieved at the beginning of the Big Bang have dissipated through billions of years, but the background radiation left over still has a temperature of 2.726 degrees Kelvin. [3] This is within 3 degrees of what physicists predicted would be left from a universal explosion, from calculations done as far back as 1948. [4]

Background Radiation. We can see evidence of the Big Bang just by turning on a television set and placing the channel selector to any unassigned channel. About 1 per cent of the visual static is residual background radiation from the Big Bang.

Scientists at Bell labs, Wilson and Penzias, discovered this effect accidentally while trying to remedy static detected by their sensitive instruments. They searched for the origin of the annoying interference, and found to their amazement the universe itself was the source, emanating a faint echo of that primordial event.

Hydrogen Depletion. Hydrogen, the lightest atomic element, is the prime component in stars. The effect of gravity causes the cores of stars to increase in temperature until the point is reached where hydrogen atoms fuse together forming helium, much as in the hydrogen bomb with its resultant release of energy.

The increasing pressure of gravity causes the helium in turn to fuse into heavier atoms, and so on, forming heavier and heavier elements. When a sufficient amount of these heavier elements has collected in the core of a star, it dies, or if it is massive enough, it explodes in a supernova. This irreversible decline of hydrogen in the universe implies that hydrogen must have been created at the beginning. [5]

Abundance of Helium. The amount of ambient helium remains virtually constant from place to place throughout the universe. [6] Peebles calculated in 1966 that the observed ratio of helium to hydrogen matches exactly what would have resulted during the first four minutes of the Big Bang using the radiation temperature published by Wilson and Penzias. [7]

Temperature Fluctuations. One of the perplexing mysteries to the Big Bang theory concerned how galaxies formed in clusters. Logically, a smooth explosion should have produced a uniform universe. In 1989, NASA launched the Cosmic Background Explorer (COBE) satellite. Its instruments detected temperature fluctuations in the cosmic microwave background amounting to barely thirty-millionths of a degree Kelvin. "The fluctuations represent tiny gravitational ripples - variations in the density of matter," *Science News* reported. The article went on to say:

> Cosmologists believe these ripples unbalanced the primordial universe enough to cause matter to begin lumping together and, after 15 billion years, evolve into the cosmic structures found today. [8]

Cosmologist Joseph Silk described the COBE findings as "the missing link." He continued:

> The lack of fluctuations has been a major obstacle in having many people accept not just [theories of] galaxy formation but the basic premises of the Big Bang. [9]

A rational conclusion is that the universe had a beginning. The time of the beginning is dependent upon the Hubble constant, the rate of expansion. This number has not been confirmed, but the Big Bang at about 12 billion years ago is a middle-ground number. Physicists in rival camps typically add or subtract up to 4 billion years. [10]

Jump Starting the Cosmos

Preston Cloud in his book, *Cosmos, Earth and Man*, said the question of first causes "transcends the bounds of science." [11] Robert Jastrow amplifies that point:

> What is the ultimate solution to the origin of the Universe? The answers provided by the astronomers are disconcerting and remarkable. Most remarkable of all is the fact that in science, as in the Bible, the world begins with an act of creation. [12]

The beginning of the universe was light. God's very brilliance shown forth and manifested itself. John Wiester narrates:

> The first particles to emerge were photons (particles of light) and neutrinos (subatomic particles that travel through solid bodies at the speed of light). These were almost instantaneously followed by electrons, positrons, protons, and neutrons. Initial temperatures were beyond comprehension, such as one hundred thousand million degrees. The Universe was filled with light. [13]

In a complicated process, physicists say, photons emanating from the Big Bang were absorbed and re-emitted rapidly, being unable to propagate freely. Ultimately, electrons and photons combined to form atoms, at which point the photons were released, and propagated as light. Nuclei were joined up with electrons to form the first atoms of hydrogen and helium. As these billions upon billions of atoms formed, and were forced out into the expanding space, gravity began to coalesce them into larger and larger clusters.

Enormous clumps of matter formed masses of clusters that grew gradually into galaxies. Gravity drew the atomic particles tighter and tighter. The heat of contraction caused the hydrogen gas to initiate a fusion process, and stars began to form. Eventually, atomic reactions took over to complete the life cycles of stars. By the same process, star formation is still taking place today.

When a star like our sun burns off its hydrogen gas, atomic particles of increasing complexity are "cooked" down. If a star is of sufficient size, it explodes at the end of its life cycle into a supernova, showering masses of debris into space. It appears our solar system was the benefactor of one or more supernova explosions some 5.5 billion years ago, that spun off a nebular cloud of basic raw materials to form our planets. [14]

After the space debris had contributed just the right mass and mix of materials we would need, the earth went through a nifty rebirthing process about 4 billion years ago. Due to gravitational compression, cosmic bombardment, and radioactive heating, the entire planet performed a meltdown so that heavier materials such as iron could sink to the core with the lighter elements rising toward the top.

Earth was blessed with a handy assortment of heavier elements we can appreciate today. An abundance of iron has afforded us a magnetic field protecting us from cosmic radiation and solar wind. A high amount of radioactive elements played a key role in remelting the earth, and it keeps our core just hot enough for a mantle with a plastic viscous upper layer. A high component of hydrous, water-forming compounds added an extensive water supply - a critical factor in supporting life. [15]

In essence, we are the benefactors of an eloquent and enormous life support system planned to exquisite and infinite detail. Ignoring any grandiose details, the biblical account of God's marvelous creation is modest in description and understated.

Genesis 1:2: "And the earth was without form and void; and darkness was upon the face of the deep."

This concise description could apply aptly to the primitive, featureless surface of our planet after it had cooled enough to allow water to condense.

Genesis 1:3-5: "And God said, Let there be light ...."

The sun, ignited first by the heat of compression, developed slowly through a hydrogen fusion process to become our source of light, heat, and energy, our material lifeline. Periods of light were named "Day," "and the darkness He called Night."

A pointing to Christ may also be seen from these passages, a "light" reflected in John 1:1-10; 3:19-21; II Cor. 4:4-6; Col. 1:12-19; Heb. 1:1-3; and Rev. 21:23.

Day Two

Genesis 1:6: "And God said, Let there be a firmament in the midst of the waters, and let it divide the waters from the waters."

The skies and seas were divided by a "firmament" boundary which God called "heaven" (Gen. 1:8). After the melting and outgassing of primordial atmosphere and water vapor, the earth was a smooth spinning ball covered by a shallow sea beneath the gaseous firmament.

Day Three

Genesis 1:9: "And God said, Let the waters under the heaven be gathered together unto one place, and let the dry land appear: and it was so."

About 3.5 billion years ago, massive dome-like blocks of granite, some over twenty-five miles in thickness, buoyed up from within the earth's mantle. They also formed massive roots, or cratons, underpinning our continental land masses today. [16]

Tremendous amounts of heat and pressure combined to bring about a separation between the relatively lighter granite land masses and the denser basalt sea basins where the waters gathered. About 2.5 billion years ago, the great vertical uplift came to an end, and the subtle lateral movement of plate tectonics took over to carry out the work of shaping the face of our planet.

Those who think the earth is young say that the effects of erosion should have worn away all the surface land completely after so many years. But just as icebergs continue to float on top of the ocean while melting, the buoyant granite continents float higher and replace the lost soil washed gradually into the sea.

According to geologists, about 200 million years ago all the earth's great continents became joined together in a giant super continent they call *Pangaea* (all lands). The sea, named *Panthalassa* (all seas), surrounded this one conglomerate land mass.

The earth has been cooling off gradually, venting steam and magma through the earth's crust in a driving motion, giving the continents a virtual shove around the globe. The fossilized remains of dinosaurs and coal seams in Alaska and Antarctica testify to their once warm-weather environs. Relief maps of the Mid-Atlantic Ridge between Europe and the Americas show the unmistakable stress of millions of years of internal pressure with resultant sea floor spreading.

Genesis 1:10: "And God called the dry land Earth ..."

Please note: God uses the word, "Earth," to denote "the dry land." To avoid the confusion caused by present-day translations, in nearly every instance for the duration of the Genesis 1-11 account, wherever the word "earth" appears we may apply God's definition, and say instead, "the dry land."

Bacteria and blue-green algae got their start over 3 billion years ago. The importance of algae early in the earth's history cannot be overemphasized. Through the process of photosynthesis, they give off oxygen, a necessary atmospheric ingredient for the more complicated life forms to follow.

The fossil record indicates that marine life, including both plant and animal life, preceded any land-based life forms. Initially, living organisms were microscopic until about 700 million years ago. Organisms of only a few centimeters in size then began to appear, including jellyfish, worms, and now extinct *tribrachidium*. [17] According to *Prehistoric Atlas*:

> The presence of fossil remains of marine animals characterises only those rocks dating back to the start of the Paleozoic Era. This proves that animal and plant life were then confined to the seas and oceans, but that all the groups of invertebrate animals alive today were already represented. [18]

Trilobites, brachiopods, sponges, and creatures of wondrous description comprised the Cambrian explosion about 543 million years ago. Biologists attribute this radical change in life forms to what might be called an "arms race." Creatures developed defensive mechanisms such as a tough shell, or were gobbled up. Fossilized shells can be found in profusion today, which mark the beginning of the Cambrian period with an exclamation point.

Strange looking colonial creatures called graptolites were prevalent, and are part of the Stomacordata group which "stands midway between invertebrates and vertebrates." [19] The first tantalizing hint at vertebrates appeared in the Ordovician period 500 million years ago such as a kind of fish without jawbones called Agnatha, forerunners to present-day lampreys.

Life on land dates to the Upper Silurian period 435 million years ago. Plants emerged from the seas, and began to colonize river banks and basins where some degree of nutritious soil was available. The first primitive fish with jawbones (Acanthodii) dates to this period. Animal life to first venture on land included earthworms, gastropods, myriapods, and arthropods, which included primitive scorpions and the precursors to insects that would be needed for pollination.

Genesis 1:11: "And God said, Let the earth bring forth grass, the herb yielding seed, and the fruit tree yielding fruit after

his kind, whose seed is in itself, upon the earth: and it was so."
Note that Genesis records land-based plant life on the third day
before marine life on the fifth day. What's going on here? Didn't
marine life precede life on land? The fossil record suggests it did,
but consistently the Bible record and the fossil record use different
start points.

Life on land, according to nature's evidence, began about 400
million years ago; ferns, club mosses, and horsetails appeared which
reproduced through spores not seeds, and were confined to moist
wetlands. The first plants with seeds date to the Devonian period
about 395 million years ago. By 355 million years ago, trees 100
feet high or more dominated much of the earth's lowlands. [20]

Conifers such as *Callixylon* began to appear, that were ancestral to pine and fir trees of today. The first blooming flowers began
to color the landscape in the Early Cretaceous some 120 million
years ago, and the earliest traces of grass date to the Upper Pale-
ocene about 62 million years ago.

Here critics point to seeming discrepancies. From the evidence
available, life in the ocean dates to even before the Cambrian period of 543 million years ago, and preceded life on land. The Bible
demonstrates consistency, though; ancient precursors to modern
men are excluded from the biblical record, and so are ancient aquatic
precursors to modern plant and animal life.

Just as primitive sea creatures preceded modern fish, likewise,
sea vegetation begat land vegetation, and all date initially to the
same period, the Ordovician. Not that it is particularly significant,
but the fossil record does indicate primitive land plants appeared
before primitive fish. So it makes no difference whether we consider primitive life or more modern life forms. The Genesis account accords either way.

"Armored fish" called placoderms date to the Devonian period
as does *Eusthenopteron* a forerunner to rhipidistians, and then cros-
sopterygians, and, perhaps, also to amphibians. The characteristics of both fish and amphibian were combined in one creature
called *Ichthyostega* also dating to the Devonian.

The latimeria and ceratodus could be called modern fish, but they appear 195 million years ago in the Jurassic period. Over 100 million years stand between seed bearing land plants and what could be called modern fish.

Genesis 1:12: What about "grass" on the third day of creation? Critics charge that grasses did not emerge until after the dinosaurs became extinct. How can 62 million year old grass predate the dinosaurs, for example, who came into existence over 200 million years ago?

Massive dinosaurs leave huge bones, which make wonderful fossils. The Cambrian explosion left a permanent record of hard-shelled marine creatures that is impossible to ignore. Any soft-shelled predecessors left scarcely a trace. The same could be said for any land-based vegetation that might have been.

Sparse fossil evidence hinders us from knowing exactly what plant life first began to grow on dry land. "Grass" may be just another instance of translation out of ignorance. The Hebrew word, *deshe,'* can mean simply "vegetation." We can verify that the earth has had land-based vegetation for over 400 million years.

Modern fruit trees, critics point out, certainly were not in existence before fishes. That seems to be true, but these verses say nothing about "modern" fruit trees. In English, fruit trees bear edible fruit; apples, pears, cherries, and so forth. The Hebrew term includes seed-bearing trees, and shade trees that do not bear edible fruit. [21]

Apple trees, for example, do not date to the Upper Silurian, but the "fruit" of any plant is its yield. Conifers were among the first land-based vegetation, and calling them "fruit trees" is consistent with the Hebrew.

Day Four

Genesis 1:14,16,17: "And God said, Let there be lights in the firmament of the heaven to divide the day from the night; and let them be for signs, and for seasons, and for days, and years ... And God made two great lights; the greater light to rule the day, and the lesser light to rule the night: He made the stars also. And God set them in the firmament of the heaven to give light upon the earth ..."

Some Bible scholars have put a strain on these passages, maintaining that the sun, moon, and stars were created on the fourth day. This is unwarranted. The emphasis in this verse is on the purpose for the heavenly bodies not their coming into existence.

If we take "the heaven" from Genesis 1:1 to include the visible universe, or cosmos, then it would incorporate the sun, moon, and stars. Even if we just take the heavens to mean "sky," it would be strangely black without sunlight, moonlight, and starlight. The *Expositor's Bible Commentary* reasons:

> So the starting point of an understanding of vv.14-18 is the view that the whole of the universe, including the sun, moon, and stars, was created "in the beginning" (v.1) and thus not on the fourth day. [22]

In the creation account, the Hebrew word *bara'* means "create," and always emanates from God. That can imply an *ex nihilo* creation, a literal out of nothing creation (Gen. 1:1), or the use of elements brought into existence previously as with primitive sea life (Gen. 1:21), also a man and his woman (Gen. 1:27). The word "made," used in Genesis 1:14-19, is the Hebrew *'asah*, a more general term, and may mean "appoint" or "accomplish" in this verse.

This distinction can be seen in Psalm 8:5: "For Thou hast made Him a little lower than the angels, and crowned Him with glory and honor." Christ was not created inferior to the angelic hosts. He was given a lowly position or status while in human form for the purpose of His earthly ministry, and thereby, was made "lower than the angels" until the resurrection.

The Septuagint avoids confusion: "God indeed made the two great luminaries, the greater luminary for the regulations of the day, and the lesser luminary, with the stars, for the regulations of the night ..."

Thus, on the first day God created the sun, moon, and stars in addition to the earth, and on the fourth day, God appointed the sun to govern the day and commissioned the moon and stars to rule the night.

Had the sun not been created until the fourth day, we would be left to wonder what caused the demarcation between the "day" and "night" named on the first day (Gen. 1:5). Furthermore, from what we know about the physics of orbital objects, it would be impossible for the earth and its sister planets to circle a blank spot in space awaiting the sun's creation.

It was pointed out in the previous chapter that a possible reason Genesis lists land plants before the luminaries began to govern is either because there were no eyes to see light, or something such as liquid water or dense clouds prevented the heavenly lights from being seen. Dense vapor clouds surrounding the primitive warm earth might not have cleared enough to enable the sun, moon, and stars to shine through, and so they could not be used for telling time.

Is it possible that cloud cover could have lasted four billion years, until after land plants appeared? Maybe not, but clouds are only a water vapor barrier which inhibit terrestrial creatures from making celestial observations. No land animals existed until the fifth day of creation.

Sea creatures also cannot make celestial observations due to a water barrier in liquid form. So for whichever reason, the presence of an obscuring barrier, or the lack of observers, the sun, moon, and stars beginning to function as timekeepers on the fourth day of creation in no way contradicts the flow of events projected by naturalists.

Day Five

Genesis 1:20, 21: "And God said, Let the waters bring forth abundantly the moving creature that hath life, and fowl that may fly above the earth in the open firmament of heaven. And God created great whales, and every living creature that moveth, which the waters brought forth abundantly, after their kind, and every winged fowl after his kind: and God saw that it was good."

Critics have a field day with these passages because birds are out of order here. Paleontologists agree unanimously that reptiles preceded birds. *Archaeopteryx*, a feathered reptile-like creature that lived 150 million years ago, is considered a likely transitional step. [23] Furthermore, whales are mammals, and necessarily follow both reptiles and birds, and should not be included with primitive sea life at all.

These are good arguments, and would seem to be valid arguments, except that Genesis was written in Hebrew, and ancient definitions differ from modern faunal classifications. A closer look at the text reveals that the order presented here should not be troublesome at all.

First of all, sharks and armored fish date to the Devonian period. When God said, "Let the waters bring forth abundantly the moving creature that hath life," these would make able representatives.

Next, the Hebrew word *'op* that has been translated "fowl," is a "flying creature," the same basic word for "insect" which probably would have been a better translation. [24] Flying insects date to 300 million years ago in the Carboniferous period, and were useful for pollinating some of the vegetation springing forth at about that time. Also, why would "fowls" be mentioned three times in three consecutive passages (Gen. 1:20-22)? If birds had been intended in all three instances it would be a curious redundancy.

Some Bible translators shun the word "whales," opting for "sea monsters," for example. This makes sense. It is doubtful that Moses ever set eyes on a whale in the Red Sea, and therefore unlikely he would have used a name for an animal he had not seen.

The English translation of the Septuagint is less confusing. "Then God said, 'Let the waters produce moving creatures having life; and winged creatures flying above the earth in the open firmament of heaven.' And it was so. God indeed made the great sea monsters and every species of moving animals which the waters produced according to their kinds, and every winged flying creature according to kind."

"Large sea creatures" satisfies the Hebrew, and one candidate is the *Dinichthys*, which could grow to over 30 feet in length, and had strong jawbones equipped with broad cutting plates making it a formidable predator. [25] Primitive amphibians began to appear in the Devonian giving way to reptiles in the Upper Carboniferous.

> Reptiles were the first vertebrates which managed to leave the aquatic environment for the laying of eggs and the development of embryos, thus opening the way to the colonisation of the subareal habitat by vertebrates. [26]

The advance of reptiles onto dry land was a big step. An amniotic egg with significantly different membranes was required than what was necessary for amphibians, which laid their eggs in water. Amphibians also lacked a waterproof skin, and became dehydrated on land. Equipped with a horny or scaly epidermis, along with a sturdier set of legs, the first reptiles such as the Cotylosaurs ventured forth on land. [27]

Now, let us see how these passages read when we make accommodations for the Hebrew after applying what we know about nature.

Genesis 1:20-21: "And God said, Let the waters bring forth abundantly the moving creature that hath life [fish], and fowl [flying insects] that may fly above the earth in the open firmament of heaven [sky]. And God created great whales [large sea creatures], and every living creature that moveth, which the waters brought forth abundantly after their kind [amphibians and reptiles], and every winged fowl [birds] after his kind: and God saw that it was good."

Day Six

Genesis 1:24: "And God said, Let the earth bring forth the living creature after his kind, cattle, and creeping thing, and beast of the earth after his kind: and it was so."

On the sixth creation day, God lets the earth bring forth creatures according to their own kind, "cattle," "creeping things," and
"beasts of the earth." We could think of these groups as domestic
animals such as livestock, wild herbivores, and wild carnivores.
"Creeping things," browsers and grazers are preyed upon by meat
eating "beasts of the earth."

Some consider "creeping things" to be reptilian, and there are
some reasons for that - none good enough, though, to rearrange the
fossil record. Psalm 148:10 categorizes "beasts," "cattle," "creeping things," and "flying fowl;" so reptiles might fit, but in Hosea
2:18, the divisions are: "beasts of the field," "fowls of heaven,"
and "creeping things of the ground," thereby placing all undomesticated animals in the "creeping thing" group.

Genesis 1:26 puts all undomesticated land animals into the
creeping category when man is given dominion over "fish," "fowl,"
"cattle," and "every creeping thing." In Leviticus 11:21-22, a "flying creeping thing" refers to insects, and names "locusts," "grass-
hoppers," and "beetles" as examples.

The "weasel," "mouse," and "tortoise" are creeping things in
Leviticus 11:29, lumping together rodents and reptiles. In Genesis
8:19, we encounter "every beast," "every creeping thing," "every
fowl," and "whatsoever creepeth upon the earth," implying two
separate categories of creepers.

The Bible did not intend to give us a neat, precisely definable
term here, and that only underscores what has been said earlier
concerning a living Bible. There is purposeful agility built into the
living Word, affording breathing space for the Bible to adapt as we
learn. That should keep us from making erroneous dogmatic assertions which have to be amended or retracted at a later date when
we know better.

With reptiles preceding birds on the fifth day in Genesis 1:21,
there is no need for reptiles on the sixth day in Genesis 1:24. Were
the "kinds," referred to in these verses, specially created with no
transitional steps in between? Not likely, but transitional life forms,
if they did exist, do no violence to the literal meanings of these
verses. Forming animals "out of the ground" does not come until
chapter 2 of Genesis, and it may be that Adam and the animals God

formed expressly for Adam's garden were formed out of the dust, or the animals Adam named could have been a subset of the animals God created in Genesis 1.

Genesis 1:26: "And God said, Let us make man in our image..."

The term "in our image" is more apt to apply to Adam specifically (more of that in chapter 18). Nevertheless, man, as creation's pinnacle, fits everybody's theory as the last to appear on earth. *Homo sapiens* have not developed into more than one species in 100,000 years, whereas many other "kinds" of animals have branched into numerous distinct and extinct species. Evidently, all the other animals have roamed the earth longer than man has, which puts science and Genesis in agreement.

Genesis 1:31: "And God saw every thing that he had made, and, behold, it was very good."

Thus a blessing was conferred upon the earth's resources, the life cycle, and the food chain. Everything in nature, man included, was "very good."

NOTES

[1] William J. Kaufmann, *Universe* (New York: W. H. Freeman and Company, 1985), 462.

[2] Hugh Ross, *The Fingerprint of God* (Orange: Promise Publishing Co., 1989), 79.

[3] Michael S. Turner, "Why Is the Temperature of the Universe 2.726 Kelvin?," *Science* (5 Nov. 1993), 861.

[4] Ibid., 84.

[5] John Wiester, *The Genesis Connection* (Nashville: Thomas Nelson Publishers, 1983), 20.

[6] Ross, *The Fingerprint of God*, 87.

[7] P. J. E. Peebles, "Primeval Helium Abundance and Primeval Fireball," *Physical Review Letters*, 16 (1966), 410-413.

[8] M. Stroh, "COBE Causes Big Bang in Cosmology," *Science News* (May 2, 1992), 292.

[9] Ibid., 292.

10 On October 26, 1994, NASA scientists released data obtained from the Hubble Space Telescope designed to help measure the Hubble constant, the rate at which the universe is expanding. The Hubble telescope obtained a distance of 56 million light years to the galaxy called M100, a spiral galaxy in the Virgo Cluster. A class of stars called "cephids" are used as standard candles. A comparison of the cephids in M100 with those found in more distant galaxies yielded an age for the universe of 8 to 12 billion years old. Previous research data had dated the oldest stars at 15 to 18 billion years old. Since stars cannot be older than the universe, it is clear that estimates of the age of the universe will be revised periodically as more data surfaces.

11 Preston Cloud, *Cosmos, Earth and Man: A Short Story of the Universe* (New Haven: Yale University Press, 1978), 25.

12 Robert Jastrow, *Until the Sun Dies* (New York: Warner Books, 1977), 11.

13 Wiester, *The Genesis Connection*, 37.

14 Ibid., 37-44.

15 Ibid., 37-45.

16 Ibid., 61.

17 P. Arduini and G. Teruzzi, *Prehistoric Atlas* (London: Macdonald & Co (Publishers) Ltd., 1982), 23.

18 Ibid., 26.

19 Ibid., 28.

20 From the exhibition on early plant life at the Smithsonian Institute in Washington, D.C.

21 Umberto Cassuto, *A Commentary on the Book of Genesis* (Jerusalem: The Magnes Press, 1944), 40.

22 Walter C. Kaiser Jr., Bruce K. Waltke and Ralph H. Alexander, eds., *The Expositor's Bible Commentary* (Grand Rapids: Zondervan Publishing House, 1990), 33-34.

23 Carl Zimmer, "Ruffled Feathers," *Discover* (May 1992), 44-54.

24 Gleason Archer, *Encyclopedia of Bible Difficulties* (Grand Rapids: The Zondervan Corporation, 1982), 59-60.

25 Arduini and Teruzzi, *Prehistoric Atlas*, 37.

26 Ibid., 32.

27 Ibid., 42.

Chapter 10

A PLACE IN HISTORY:
ADAM AND ASSOCIATES

Daniel's curiosity was answered, "But thou, O Daniel, shut up the words, and seal the book, even to the end; many shall run to and fro, and knowledge shall be increased" (Dan. 12:4). The words the Lord had directed Daniel to write were for another age. What those words meant was none of his business. So caution flags should be flying whenever one attempts to interpret or reinterpret Scripture.

The time may not have arrived for the complete meanings to be known. We have more knowledge yet to acquire. But we will continue to gain knowledge, and we must find suitable answers eventually lest we blunder into, or even beyond, the 21st century. Appropriate answers are essential; we are still burdened with Bible interpretations stuck in a time when people thought the world was flat, and the sun, moon, and stars were in orbit around our pizza-shaped planet.

If we believe the Bible is trustworthy in recording history, then Adam and Eve were *de facto* historical figures, not symbolic representations concocted by Moses or some other source. Establishing Adam's approximate historical time frame is essential if we are to understand the origins of man fully from the standpoints of what we know historically and anthropologically, as well as from what we are told biblically.

In writing Luke and Acts, the writer, Luke, incorporated small amounts of secular history as well. As a result, readers many centuries removed have minimal trouble determining when the reported events took place. In the beginning of Genesis also, sufficient peripheral information is recorded to give us a fairly accurate historical perspective.

What has impaired Bible interpreters from the start is a propensity to conclude that the Hebrew text somehow designates Adam at the apex of our species. Most New Testament manuscripts (excepting the Vatican and at least three others) position Adam as the first "man" (I Cor. 15:45), but what definition is to be applied? Could Adam have been the first hominoid, the first hominid, first of the genus *Homo*, first of the *Homo sapiens*, the first Caucasian, or the first of a Near East people from which present-day Jews, Arabs, and some others have derived? Remember, we have only one Adam, and he lived only once.

Trying to establish a date in history where Adam could have started the human race is a futile exercise. Regardless of whether the Australopithicines are in our line of ancestry or not (a subject of debate), placing Adam at the start of the genus *Homo* would mean that he must have lived over 2 million years ago without removing the possibility that he could have been in the company of remnant Australopithicines. [1] *Homo erectus*, dating from 1.6 million to 300,000 years ago, had the knowledge of fire and used stone tools. But were they ancestors, descendants, or merely prototypes? [2]

Placing Adam at the start of the modern *Homo sapiens* solves nothing; he would have been overlapped fore and aft by both archaic *Homo sapiens* and Neanderthals, which appear in the fossil record earlier than modern humans, and were still living 65,000 to 70,000 years after modern humans began. [3] So it is not feasible that Adam could have been alone in the world from the outset no matter when he appeared, if it was within the last million years.

The Bible is too specific in detailing the culture of Adam's day to allow us to propel him back in time. Also, the genealogies in Genesis 5 and 11 are a prohibitive factor. But even if we could slide Adam back tens of thousands of years so that he could start the *Homo sapiens* species, or even one of the races, there is no place in the history of hominids where Adam could have commenced his line of descendants in isolation. By all indications, Adam entered a populated world.

FORERUNNERS TO MODERN MAN

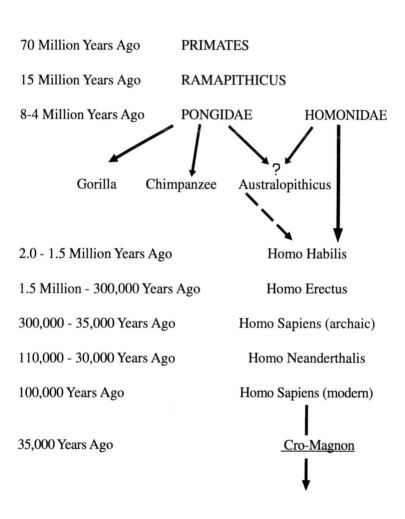

70 Million Years Ago	PRIMATES
15 Million Years Ago	RAMAPITHICUS
8-4 Million Years Ago	PONGIDAE HOMONIDAE

Gorilla   Chimpanzee   Australopithicus

2.0 - 1.5 Million Years Ago	Homo Habilis
1.5 Million - 300,000 Years Ago	Homo Erectus
300,000 - 35,000 Years Ago	Homo Sapiens (archaic)
110,000 - 30,000 Years Ago	Homo Neanderthalis
100,000 Years Ago	Homo Sapiens (modern)
35,000 Years Ago	Cro-Magnon

DESCENT OF MAN

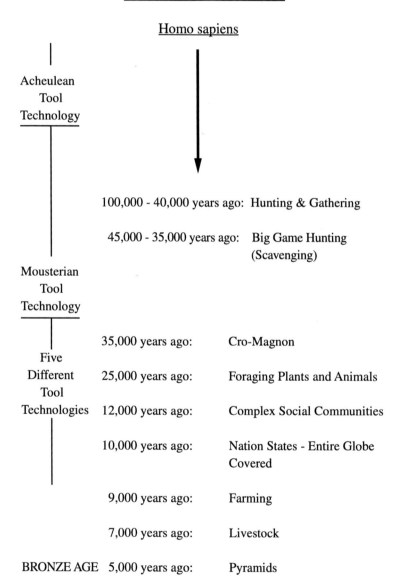

Homo sapiens

Acheulean
Tool
Technology

100,000 - 40,000 years ago: Hunting & Gathering

45,000 - 35,000 years ago: Big Game Hunting
                             (Scavenging)

Mousterian
Tool
Technology

35,000 years ago: Cro-Magnon

Five
Different   25,000 years ago: Foraging Plants and Animals
Tool
Technologies   12,000 years ago: Complex Social Communities

                10,000 years ago: Nation States - Entire Globe
                                         Covered

                9,000 years ago: Farming

                7,000 years ago: Livestock

BRONZE AGE 5,000 years ago: Pyramids

Mitochondrial Eve

The "Eve hypothesis" was developed from pioneering work in mitochondrial DNA (mtDNA) published by Wilson and Sarich in 1987. According to them and other recent researchers, there is evidence that all human beings have descended from one common female genotype who lived in Africa about 200,000 to 100,000 years ago.

In addition to nuclear DNA which contains our individual genetic code, each cell of our body contains mtDNA also. These minuscule structures within each cell generate its energy, and come initially from the egg of the mother. Researchers build ancestral trees by comparing the amount of divergence in the mtDNA code. The data suggests the roots of that genetic tree are in Africa. [4]

Support for the "out of Africa" genetic model can be derived from the morphological diversity seen among black Africans today. Ancient populations originated in Africa, according to this theory, since a greater amount of time is required to accumulate a larger number of mutations resulting in greater genetic distance between isolated populations.

Contrast the morphological distinctions between Pygmies, Hottentots, and Bantus, for example, with the relatively homogeneous morphology of the various tribes of native Americans. The greater morphological differences correlate with greater genetic distance, and leads to the conclusion that more time stands between surviving populations of black Africans and their common ancestors than between surviving populations of native Americans and their common source. The African populations, therefore, are older and originated earlier.

Researchers at the Natural History Museum in London prefer the "out of Africa" model. It is believed that only there *Homo erectus* gave rise to modern humans. These spread throughout Europe and Asia, and displaced whatever remnant hominid populations they may have encountered in their migrations.

A number of distinguished paleontologists disagree, and they have published data suggesting a commingling between ancient and more modern peoples. Their evidence supports "regional continuity," they maintain, meaning that local populations of archaic types begat modern types eventually in more than one location.

An analysis of human fossils found in Israel and Africa, when compared with older *Homo erectus* remains, led researchers to place *Homo erectus* directly in the line of hominids that culminated in modern man. *Science* reported:

> These modern-looking fossils all date to about 100,000 years and appear at the end of a sequence of fossils that stretches back to 400,000 years ago, which seem to show a gradual transition from their *Homo erectus*-type forebears to early modern humans. [5]

What unity there is among contending parties was summed up:

> In spite of the contention, all parties can agree on one thing. The proto-human fossil record begins in Africa, with a species now called *Homo erectus*. After evolving in an African homeland, all concur, *Homo erectus* migrated to Europe and Asia about 1 million years ago. But after that, comes the Great Divide in paleoanthropology. [6]

Although two theories are competing for prominence, what has been generally agreed upon by both molecular biologists and paleoanthropologists is that all humans are biologically connected as evidenced by our DNA signatures [7] (and confirmed in Acts 17:26). When and under what circumstances ancient "Eve" got here is still an open question, or even if there was more than one "Eve."

The temptation among some Bible apologists has been to speculate that Adam lived at a similar early date, say, 100,000 years ago or more, and the origins issue is seemingly resolved. Even if the Bible was accommodating (and it isn't) how would we explain the various precursors predating that point in history? Can they just be swept under the rug?

A Time for Adam

According to Genesis, Adam was the first to have a covenant relationship with the Creator, the first to be accountable, the first to suffer the consequences of sin, and the first in the line of descent leading to the Savior. That does not mean necessarily, however, that Adam was the first biped with an opposable thumb and a cranial capacity of 1300 to 1400 cubic centimeters.

The task of finding some place to inject Adam into human history can be simplified if we let the Bible do the talking. What we find is that the inspired text does not place Adam at the head of our species. References to tents, farming, and raising livestock (Gen. 4:2, 20) suggest that Adam was not surrounded by cave-dwelling hunter-gatherers, an occupation of our quite ancient forefathers.

Archaeologists place the beginnings of modern man at 10,000 years ago with the advent of farming techniques. [8] Adam's placement at roughly 7,000 years ago from the Genesis genealogies, coupled with the mention of farming in the Genesis text, makes this a compatible time frame. This puts Adam in relatively recent history not ancient history.

Tubal-cain worked with "brass and iron" (Gen. 4:22). The Hebrew word for "brass" also means "copper," and copper tools have not been found dated any earlier than 10,000 years ago. Although iron smelting would be out of the question at that early date, there is evidence that bog iron was beaten into rudimentary tools, and iron was known as far back as 4000 BC. [9] Or what may have looked like iron to the ancients, could have been tin. Copper and tin together make bronze, and the Bronze Age is identifiable in history, starting about 3000 BC. [10]

Stone tools would have been of little use to Noah when he needed to construct a massive watertight ark. Metal tools suitable to build an ark would have been available only if the pre-flood patriarchs lived in the period of what archaeologists call modern man; that is, after 10,000 years ago. The old Stone Age periods may not have passed completely by Adam's day, but human history was certainly into the Bronze Age by the time of Noah. These are only a few of the reasons the first chapters of Genesis cannot be relegated to the distant past.

Knowing that Adam was in the stream of humanity, rather than at the start, precludes his being ancestral to most of us. Of course, a date such as 100,000 years ago for the emergence of modern-looking human beings may undergo revision in the future, but barring any drastic changes, there is no comfortable niche for Adam any time before communicative bipedal creatures of some description had already commenced on planet Earth. What became of them is the real issue.

These creatures either died out, leaving the world devoid of humanity until Adam was created, or else they left progeny which can be seen riding subways, doing stand-up comedy, and raising gifted children today. In other words, they are us. Considering the likelihood that modern man has ancient ancestors, Adam was either nonexistent - a notion the Bible rejects - or else he was inserted, so to speak, into the train of humanity. This latter alternative is the solution we will explore.

A Concerned Cain

Cain's lament in Genesis 4:13-14 weights the issue heavily as to whether Adam had company or not. The covenant family was reduced by 25% when Abel was murdered; only Cain and his parents were left. Cain's first words upon hearing the Lord's punishment and upon God's banishing him from Eden were out of fear that someone would kill him. Is it likely that his immediate worry would have been of being sought after and killed by future unseen and unknown generations from Adam? Cain had a whole world in which to hide.

In a human-free environment, the threat of isolation and being alone in the world would have been a natural fear. Cain might have been concerned about wild animals attacking and eating him, but he did not register any fear about that possibility. His only concern was that someone would end his life just as he had slain his own brother.

God answered Cain's plea by providing a sign for him (Gen. 4:15). Cain's anxieties were justified because the Lord took positive action to quiet his fears. We have no way of knowing what that sign or mark was, but evidently it was necessary. There must have been potentially hostile tribes of men in the vicinity. Cain was aware of it, and the Lord's action attested to his justifiable fear.

Removing the Shackles of Prejudgment

Once we hold up to scrutiny the common misconception that Adam was the first human, and consider the probability that other human beings were already living in Adam's proximity, previous pitfalls in the Genesis narrative disappear. Passages that had obscure meanings now take on dimension.

For example, the "Nephilim" or "giants" (depending on translation) in Genesis 6:4 are now identifiable as prehistoric, or pre-Adamic - not in Adam's line of descendants or ancestry. If we can shed our unwarranted prejudices, admittedly a difficult task, we may then look at Genesis afresh.

Keep in mind, the details in the Genesis narrative are sketchy at best. Paleontologists are not in agreement over the exact course of man's descent due to incomplete fossil evidence of early hominids; and it is still too soon for gene research to give us a definitive picture. Nevertheless, if we cast off the shackles of prejudice, we may be able to examine the Genesis text with a view toward what may not be entirely provable, but is certainly possible, plausible, and, if I may be so bold, indeed probable.

The Image of God

Genesis 1:26, 27: "And God said, Let us make man in our image, after our likeness: and let them have dominion over the fish of the sea, and over the fowl of the air, and over the cattle and over all the earth and over every creeping thing that creepeth upon the earth.

So God created man in His own image, in the image of God created He him; male and female created He them."

What does it mean to be created in God's image? Humbert raised the possibility that man was given the same "physical outward appearance" as the deity. [11] Our physical environment requires a certain functionality of our physical bodies, however, that would not be imposed upon a Creator-God.

"The ancient orient shows us with ever increasing clarity that the purpose and function of an image consists in representing someone," according to Edmond Jacob. "An image, that is to say a statue of a god is the real presence of this god ..." [12] In that context, Adam would have been God's representative to the world, or conversely, the world's representative to God, but in either case, an already populated world. In M'Causland's words:

> Adam then Appears in the majesty of God's likeness, ushered into the world in the fullness of time, to draw his fellow-creatures to the development of the hidden treasures of wisdom and knowledge in their widest and deepest sense. With other races of human beings surrounding him, he is a more perfect type of the second Adam, than if he had been a solitary individual occupying the wide domain of the habitable earth, without a fellow-creature to behold him a being made in the likeness of the Creator. [13]

By the phrase, "the image of God," the writer of Genesis may have been alluding to the inner essence of us which is an integral part and yet unseen - our soul, or our spirit. That may not have been an altogether uniquely Adamic feature. Even though Adam was infused with something which gave him a kind of kinship with the deity, we are in the dark with respect to Adam's neighbors.

It should be pointed out that to the Hebrews, body and soul were a single entity, not subdividable into separate components. In our culture, we prefer to differentiate, separating the physical body from the spiritual soul. Whether those living at or before the time of Adam had "souls" capable of salvation, or punishment, is beyond the realm of our present-day understanding or the scope of this inquiry. Let future generations ponder that one.

Who is the "man" in Genesis 1:27? It has been argued that this verse applies to generic man, all *Homo sapiens,* and not exclusively to Adam and his following generations. But most Bible scholars believe this passage applies solely to Adam and Eve, and their descendants who came under the Adamic covenant. This is the preferred view, and implied in Genesis 5:1-3:

"This is the book of the generations of Adam. In the day that God created man, in the likeness of God made he him; male and female created he them; and blessed them, and called their name Adam, in the day when they were created. And Adam lived an hundred and thirty years, and begat a son in his own likeness, after his image; and called his name Seth."

Who was created "in the likeness of God"? The man, Adam, who "lived an hundred and thirty years, and begat a son," "and called his name Seth." Who were not created "in the likeness of God?" Those who did not live "an hundred and thirty years," and did not "begat a son" called Seth - the indigenous populations.

It is also true that traditionally the majority of Bible scholars have thought all of humanity was created in the image of God because supposedly all of humanity started with Adam. This stand has been taken, however, with a certain nonchalance not only for the fossil record and the genetic evidence, but also without heeding the qualifiers in the Scriptures themselves. Adam was created, and then Eve, but it is unfounded to think ancient precursors are encompassed by Genesis 1:27.

It is significant that two different Hebrew words for "man" are used in the Genesis text, *'adam* and *'ish.* M'Causland reasons why:

> The words "Adam" and "ish" are clearly different in meaning; and to use them indiscriminately, as having the same signification, tends obviously to obscure the true import and significance of the Scripture text. Had the translation been literal, the sense of the sacred record would have been more readily discovered, and the reader would recognize at a glance, that the history which he has conceived to be a history of the origin of all mankind, is simply a

> record of the creation of "*the Adam*," the last, and not the
> first of created men, and a history of his lineal descen-
> dants. [14]

Adam, as God's chosen, was the first man capable of achiev-
ing God's kingdom, and that was passed to his generations until
Christ's sacrifice at the cross changed the equation and brought a
new covenant. Presumably, any outsiders living at the time of Adam
would have been outside the old covenant, and unable to enjoy this
unique status, which included the hope of being claimed by God
through (1) the Adamic bloodline, (2) the discipline of self righteousness, and (3) the ritual of animal sacrifice.

The beginnings of God-awareness or seeking after God can be
substantiated in history by the evidence of religious relics and al-
tars dating as far back as 24,000 years ago, [15] but there is no evidence that the Creator manifested Himself to any of these forerunners as He did to Adam.

Catal Huyuk in south-central Turkey was excavated in the
1960's. This city was settled as far back as possibly 8300 BC, but
by about 5600 BC it was abandoned. [16] From analysis of skeletal
remains found there, a French expert concluded that two distinct
racial types were represented, one European, the other Asian. [17]
Although many shrines were unearthed at Catal Huyuk, there were
no signs of animal sacrifice.

> ... animal sacrifice apparently was not practiced inside
> the shrines, as there is no evidence of a slaughtering block
> or a catchment for the runoff of blood. [18]

If animal sacrifice as a covering for sin began with Adam and
his descendants after the Fall, then apparently Catal Huyuk was
not populated by Adamic or Semitic populations. Also, 5600 BC
is far too soon for any Semites and a little too soon for Adamites.
Thus, Catal Huyuk must have been a pre-Adamic city, and the residents there were not in "the image of God."

Relevant Time Periods
In The Near East And Mesopotamia

Paleolithic	20,000 to 10,000 Years Ago
Natufian	10,000 to 8000 BC
Neolithic	8000 to 5000 BC
Hassuna	6000 to 5250 BC
Samarra, Halaf, Eridu	5500 to 4700 BC
Ubaid	4500 to 3500 BC
Uruk	3500 to 3100 BC
Jemdet Nasr	3100 to 2900 BC
Early Dynastic	2900 to 2370 BC
E. D. I	2900 to 2750 BC
E. D. II	2750 to 2600 BC
E. D. III	2600 to 2370 BC
Sumer and Accad	2500 to 2000 BC
Old Babylonian, Larsa	2000 to 1600 BC
Cassite, early Assyrian	1600 to 911 BC
Assyrian	911 to 612 BC
Neo-Babylonian	625 to 539 BC

A Place for Adam

As high school students of world history are taught, the Fertile Crescent where civilization began is in the ancient Near East. Identifying the various cultures that have flourished in that region has been done with meticulous care made possible by years of compiling archaeological data. The earliest identifiable people belong to the Neolithic Natufian culture which was spread from Palestine to Syria, and date to about 12,500 to 10,500 years ago, clearly a pre-Adamic date. The oldest city identified with Natufian culture was Jericho. [19]

> Contrasts among Jericho, Catal Huyuk, Jarmo, and Umm Dabaghiyah - all about 6000 BC - suggest a considerable regionalization within widely scattered Neolithic communities of the Near East. [20]

From the mound of Tell Hassuna in northwestern Iraq, the Hassuna culture takes its name, and dates to 6000 to 5250 BC. Numerous agricultural villages have been unearthed in Iran, Turkey, and Palestine that were contemporaneous with the Hassuna.

The Hassuna, identified by their coarse pottery wares, were replaced gradually by the Samarra culture starting about 5500 BC. At Tell-es Sawaan in Iraq, alabaster female figurines were discovered along with ornaments of turquoise, carnelian, greenstone, and copper. The presence of widely disparate materials in one location indicates trading practices, and that trade routes had been established by that time. [21]

Dating to 5500 to 4700 BC, the Halaf culture succeeded, but overlapped the Samarran. Halafian ceramics have been discovered from the Mediterranean coast to Iran, though the Tigris-Euphrates region south of Baghdad may have been uninhabited at this early date.

From similarities in pottery shards and other artifacts, the highly developed Sumerian, Babylonian, and Assyrian civilizations that flourished in the third and second millennium periods can be traced to the late Neolithic villages of around 5500 BC. There is no break that one would expect to see if there had been a catastrophic termination of mankind and a subsequent renewal, something that is popular among "gap" proponents.

The Bible describes Southern Mesopotamia clearly as the place of origin for Adam and his generations. The rivers, Hiddekel (Tigris) and Euphrates, the cities of Erech and Ur all point to this region - a region that came to be called "Sumer."

Located four miles from the ancient Sumerian city of Ur is the small archaeological mound of al-'Ubaid. The settlements in Southern Mesopotamia dating from 4500 to 3500 BC are assigned collectively to the Ubaid culture. Whether or not pre-Ubaid sites exist in Southern Mesopotamia is a subject of controversy. Some archaeologists believe that fluctuations in the level of the Persian Gulf may have erased any traces of earlier settlements.

The origin of the Ubaid culture is unknown. The Halafians were flourishing in the north at about the same time Ubaidan farmers began to settle the southern delta of the Tigris and Euphrates rivers. The climatic conditions seem unlikely for a garden of Eden until the advancement of irrigation could bring water to the area. Irrigation technology began to be employed during the Ubaid period.

By 3500 BC, the Ubaidans were living in townships from Mesopotamia to Syria to Turkey. The subsequent flood at the time of Noah could have wiped out the Ubaidans, although there is some evidence the Sumerian culture may have derived from the Ubaidan. Broken pieces of pottery show subtle transition from Ubaid ware to Uruk ware. This is more indicative of gradual change through the influence of friendly contact with neighboring cultures, than it is of a foreign invasion and replacement by conquest. Yet some archaeologists prefer the displacement model, and believe the Sumerians were a discrete population.

The purpose of designating these ancient populations as Halafian, Ubaidan, or Sumerian is primarily to place them in time and place context, and need not necessarily imply ethnic differences. The flood must have devastated Southern Mesopotamia leaving behind ruined cities which the next generations of Sumerians could repopulate and build upon. Whether Ubaidan fathers had Sumerian sons is unknown.

When it comes to identifying candidates who may have enjoyed the Tigris and Euphrates region prior to Adam's creation, there are two or three choices depending on the precise date of Adam's arrival. We can select the earlier Halafians, the Ubaidans, or the later Sumerians, although the Ubaidans seem the most likely:

> About 4500 BC the region was settled by people who came to be called Ubaidans. They in fact settled most of the sites where the great cities of Sumeria [Sumer] were to grow - including Ur (where Wooley found their remains under the silt of the flood). Later they spread up the valley, succeeding the Halafians and becoming the first people to dominate the whole of Mesopotamia. [22]

The harsh, arid conditions might have caused the Halafians to make only brief appearances in the south, or maybe they never got there at all. Ubaidan pottery has been found at the lowest levels of excavated cities in Southern Mesopotamia, making them the earliest inhabitants that can be identified as founders of the region, succeeded by the Sumerians. It is likely that Adam and his generations were surrounded from the beginning, or became surrounded by first Ubaidan, and then Sumerian culture.

Irrigating the Garden

Genesis 2:5-6: "And every plant of the field before it was in the earth, and every herb of the field before it grew: for the Lord God had not caused it to rain upon the earth, and there was not a man to till the ground. But there went up a mist from the earth, and watered the whole face of the ground."

This is a useful passage for demonstrating that Bible interpretations which exclude pertinent extra-biblical data can produce dubious opinions and perplexing conclusions. From this verse, Henry Morris argues for a "vapor canopy" over the early earth, and reasons:

In the original world, however, there was no rainfall on
the earth. As originally created, the earth's daily water
supply came primarily from local evaporation and condensation. [23]

Morris reaches this conclusion based solely on his reading of
the biblical text, deducing that rain doesn't come until the flood,
notwithstanding that no one has discovered any place in the world
where mist or fog oozes naturally out of the ground in sufficient
volume to water humans, livestock, and crops. We would also be
left to wonder what furnished the rivers in Genesis 2:10-14 with
water. Were the Tigris and Euphrates not supplied by snow melt
and rainfall as they are today?

In their *Commentary On The Old Testament*, Keil and Delitzsch
explain Genesis 2:5 as follows:

The creation of the plants is not alluded to here at all, but
simply the planting of the garden in Eden.

They too slide down the slippery slope to a woeful opinion.
"This was dependent upon rain," they decide, and conclude that
the mist or vapor in Genesis 2:6 was the "creative beginning of the
rain itself ..." [24] So even though the Bible states in the previous
verse "for the Lord God had not caused it to rain," nevertheless,
rain it was, according to this respected Bible commentary.

So which is it, rain or no rain? Will these passages be under-
stood when "knowledge shall be increased," as it says in Daniel?
Let's try it, and see. The following is taken from the *Cambridge
Encyclopedia of Archaeology* pertaining to ancient Mesopotamia:

The culmination of these prehistoric advances is to be
found in the 'Ubaid period of the sixth and fifth millen-
nia, when the earliest settlements are known from Sumer.
This area was characterized by the very great fertility of
its alluvial soil and - outside local areas of marsh and
lagoon where a specialized fishing, hunting and

collecting economy could have been practised - an extremely arid environment that necessitated the use of irrigation for successful agriculture. 25

Could "an extremely arid environment" be described as a place where the "Lord God had not caused it to rain"? Could a "mist from the earth" that "watered the whole face of the ground" refer to a land "that necessitated the use of irrigation for successful agriculture"?
Even before the first cities began to appear on the Mesopotamian plain, sizeable settlements were being supplied by irrigation.

> The biblical city of Jericho, a center for salt trade, flourished during the seventh millennium BC in the desert near the north end of the Dead Sea. Water diverted from a spring nourished its fields. 26

Driver suggests Genesis 2:5-6 is about irrigation:

> Provision made for the irrigation of the garden. The reference is implicitly to a system of canals, such as existed in Babylonia ... 27

The Septuagint offers further assistance. In the Greek text, the word is not "mist," but "fountain." The RSV uses "stream." Could part of an irrigation system be called a "fountain"? Could a canal be called a "stream"? At least could we agree that the words fountain and stream better describe a system of irrigation than they do a vapor canopy? It seems "there was not a man to till the ground" for an uncomplicated reason. No one had irrigated the desert soil; thus no plowing had been done, so no crops could be grown.
Genesis 2:7: "And the Lord God formed man of the dust of the ground ..." The Hebrew word appearing as "man" in this text is 'adam. Was the "man" formed out of the dust the father of our species, or was it Adam, father of Cain, Abel, and Seth?
Genesis 2:8-10: "And the Lord God planted a garden eastward in Eden; and there He put the man whom He had formed.

And out of the ground made the Lord God to grow every tree that is pleasant to the sight, and good for food; the tree of life also in the midst of the garden, and the tree of knowledge of good and evil.

And a river went out of Eden to water the garden ..."

It is unlikely that a river, synonymous with "brook" or "creek," is intended. Water falls on the ground, trickles into streams, and flows to rivers, which empty in the sea - the exact opposite of what the verse states. The purpose of irrigation canals is to carry water from the rivers to the ground - precisely what the verse states. There were no "rivers" in Babylon (Psa. 137:1), only canals. Since we are afforded a clear example in Psalm 137 and Ezek. 1:3; 3:15, 23 where "rivers" means canals, a basis exists to use the same definition in Genesis. In other words, there was a place called Eden, out of which a canal ran eastward to irrigate the garden, where God placed Adam.

We know that Southern Mesopotamia was laced with a canal network, the remains of which can still be seen today as lines in the desert. Obviously canals required people to dig and maintain them. What cries out for attention, though, is this: How could Eden be identified and named as a place distinct from the garden if there was no citizenry?

Take any place - London, England, for example. Was there ever a time when London was unoccupied? Well, yes, but no one could have called it "London" then. The principle is the same concerning Eden. Isaiah speaks of the Lord making the wilderness of Zion "like Eden" (Isa.51:3). Apparently Eden was a place for people, and had to have people before it could be called "Eden."

Locating the Garden

Four rivers are named which fairly delineate the approximate location of Adam's home.

Genesis 2:11-14: "The name of the first is Pishon: that is it which compasseth the whole land of Havilah, where there is gold; and the gold of that land is good: there is bdellium and the onyx stone. And the name of the second river is Gihon: the same is it that compasseth the whole land of Cush." (Because *cush* also means "black," translators guessed at "Ethiopia." This is in some translations.) "And the name of the third is Hiddekel: that is it which goeth toward the east of Assyria. And the fourth river is Euphrates."

Although one could get the impression that one river separates into four, "and from thence it was parted, and became into four heads" (Gen. 2:10), it can also be interpreted that four rivers become one, a confluence of rivers, which better suits the topography of Mesopotamia and the nature of rivers.

> ... the term "heads" can have nothing to do with streams into which the river breaks up after it leaves Eden, but designates instead four separate branches which have merged within Eden. 28

The fourth river is easiest to identify as the well-known Euphrates, and it is joined by three other rivers before emptying into the Persian Gulf. The Hiddekel is the Tigris, the "great river" Daniel stood beside (Dan. 10:4). It originates in the region of Assyria, flowing southeast until it joins the Euphrates at a point east of Assyria, just as stated in the Bible.

M'Causland identifies the Gihon as the "Gyudes" of the ancients, 29 the modern Karkheh joined by the Kashkan river in the region of Cush, or Kush, in Eastern Mesopotamia, later called Khuzistan. Pishon may be the ancient "Pasitigris," the modern Karun which joins the Euphrates just above the Persian Gulf. This matches roughly with Driver's assessment. Conceding that more than one place may have been called "Havilah," Driver places it "most probably" in the northeast of Arabia on the west coast of the Persian Gulf: "The gold of Arabia was famed in antiquity." 30

Put in perspective, the most ancient cities of Southern Mesopotamia, Eridu and Ur (discussed in following chapters) were located about 85 miles west of the junction of the Tigris and Euphrates, and Eridu was furnished water via canal from the Euphrates.

Enuma Elish, Early Creation Epic

The first post-flood people who can be identified historically as likely descendants of Adam are the Semitic Accadians. Most authors on this subject believe it was an influx of Semites (See note [31]) from the early third millennium BC that the Sumerians knew as "Martu." [32] The Accadians learned their writing skills from the Sumerians, apparently, and began to record their own versions of history in their own language using the same cuneiform technique.

Enuma Elish is one of the early creation epics written in Accadian or Babylonian cuneiform. It has been compiled from tablets found at Ninevah, Ashur, and Kish. [33] According to legend, father Ea, second in the early Accadian trinity, begat the heroic Marduk who slays the rebellious Tiamat. (For a shadow of this see Psa. 89:9, 10 and Isa. 51:9.) Thereupon:

> *He split her like a shellfish into two parts:*
> *Half of her he set up and cield it as sky ...* [34]

The one who "contrived the uprising" was the evil Tiamat's commander-in-chief, Kingu:

> *They bound him, holding him before Ea.*
> *They imposed on him his guilt and severed his blood*
> *   (vessels).*
> *Out of his blood they fashioned mankind.*
> *He imposed the service and let free the gods.*
> *After Ea, the wise, had created mankind,*
> *He imposed upon it the service of the gods.* [35]

In this account, the blood of Kingu was used, but in another legend the blood is mixed with clay. [36] Although somewhat gory in describing the mode of their creation, the Accadians also seemed to be aware they were not alone in the world. In Accadian literature, frequent references are made to the "black-headed" people, as in this verse: "May his words endure, not to be forgotten, in the mouth of the black-headed, whom his hands have created." [37]

The "black-headed" refers to the Sumerians who supplanted the Ubaidans, or conceivably, it could be a reference to some other race of people. But whoever they were, they were not Semites (or Adamites) judging from Accadian poetry:

> *May he shepherd the black-headed ones, his creatures.*
> *To the end of days, without forgetting, let them acclaim*
>   *his ways.*
> *May he establish for his fathers the great food-offerings;*
> *Their support they shall furnish, shall tend their sanctu-*
>   *aries.*
> *May he cause incense to be smelled,...their spells,*
> *A likeness on earth of what he has wrought in heaven.*
> *May he order the black-headed to re[vere him],*
> *May the subjects ever bear in mind their god,*
> *And may they at his word pay heed to the goddess.*
> *May food-offerings be borne for their gods and goddesses.*
> *Without fail let them support their gods!*
> *Their lands let them improve, build their shrines,*
> *Let the black-headed wait on their gods.*
> *As for us, by however many names we pronounce, He is*
>   *our God!* [38]

The Semitic Accadians evidently considered the "black-headed" a separate people, racially distinct, and polytheistic regarding religion. The light-skinned, dark-haired Sumerians best fit this description, and they spoke an unrelated language. Incidentally, we are speaking of a time long before the tower of Babel incident.

Early Adamite populations must have lived in relative isolation at the beginning since they developed a language entirely unlike the Sumerian language. But by the time the Sumerians had learned to write, some of the earliest names recorded are Semite (or Adamite), demonstrating the close contact between these two cultures very early on.

A Warning

As we all know, Adam encountered some serious trouble. The hope for eternal life does not come entirely without strings attached, but for Adam the commandment was simple.

Genesis 2:17: "But of the tree of the knowledge of good and evil, thou shalt not eat of it; for in the day that thou eatest thereof thou shalt surely die."

Adam is warned that death awaits if he chooses to ignore God's command. But what manner of death? There are two different ways we could view death in this passage. First, is physical death - a death we can all understand. Did Adam die physically in the "day" that he bit into the forbidden fruit? No, he lived for 930 years. Was Adam designed such that he might have lived eternally had he obeyed? Perhaps not, the tree of life was part of the process somehow.

Physical death has been an integral part of the balance of nature since the first bacteria and blue green algae appeared. Adam must have been able to observe death in some form, or the very meaning of the word would have been ambiguous. We will see from subsequent passages that spiritual death is surely indicated. Possibly, physical death came as a further repercussion.

Animals of the Garden

Genesis 2:19: "And out of the ground the Lord God formed every beast of the field, and every fowl of the air; and brought them unto Adam to see what he would call them; and whatsoever Adam called every living creature, that was the name thereof."

The animals of the earth are created before man in Genesis 1:21-27. The reversal of the order in the second chapter of Genesis indicates possibly that the animals in Genesis 2 are a group suited to Adam's garden. Adam conferred names to those animals in his immediate locale. From Archer:

> ... God then gave Adam a major assignment in natural
> history. He was to classify every species of animal and
> bird found in the preserve. With its five mighty rivers
> and broad expanse, the garden must have had hundreds
> of species of mammal, reptile, insect, and bird, to say
> nothing of the flying insects that also are indicated by the
> basic Hebrew term '*op* ("bird") (Gen 2:19). [39]

What was required of Adam was not a complete inventory of
every forerunner to the world's present-day animal populations
covering the entire globe, but just a survey of the animals that were
in close proximity to the garden of Eden, the ones Adam would
encounter on a daily basis.

The words of Genesis 2:19 seem to describe a method of creation. There is no need to take issue with Scripture. Forming this
particular animal population "out of the ground" requires no helpful rationalization. We are not given any clues as to how numerous
this local group might have been. These animals may have been
the product of an act of special creation after the creation of Adam,
or they may have been the product of millions of years of
descendence from God's creation pronounced "very good" in Genesis 1:31. If that is the case, they would have been biologically
compatible with animals living outside the region.

A Search for a Bride?

Genesis 2:20. After naming the animals of the garden, there
was still something missing, "but for Adam there was not found an
help meet for him."

A search can be implied by the words "was not found." A
search for a helpmate to be both wife and companion would be
ridiculous if the world at that time contained only birds, beasts,
cattle, and creeping things - but what if one or more settlements of
humans was already in the vicinity?

Available females must have been living nearby, one of which
Adam could have chosen for his wife. We can deduce that from

archaeological history. From the Bible we can conclude that none was suitable, so Adam had an operation resulting in Eve. As confirmation of an act of special creation for the first covenant couple, Genesis 2:21-23 gives us a graphic description. Paul confirms this mode of origination in I Timothy 2:13.

We do not know what all the requirements were, why other females residing in neighboring villages were unsuitable. We do know Adam enjoyed long life. It is probably out of consideration for Adam's projected longevity that God gave him a wife he could live with on into his old age. Furthermore, Adam was without sin, and to be equally yoked, a sinless Adam needed a wife who also was without sin.

If a native population was in the vicinity from which Adam could have chosen a wife, it then becomes helpful if we know just where Eve did come from. Eve's origin could become a sticking point in a human-dominated environment if the details are not spelled out carefully.

God's desire was for Adam's wife to be distinctive, just as Adam was. A portion was taken from Adam's side, God fashioned it into Eve, now they both could enjoy 900 or more years of wedded bliss. We are free to speculate about the origins of Cain's wife, or Noah's wife, but not about Eve.

Darkness on the Horizon

Genesis 2:25: "And they were both naked, the man and his wife, and were not ashamed."

Thus the book is closed on what was until then a perfect relationship - man and wife living in sublime harmony with each other, with the lesser animals, with their neighbors, and with their Creator. The story might have ended there had the world been inhabited solely by material creatures, but a spirit world as part of our earthly environment complicates the matter.

"For we wrestle not against flesh and blood, but against principalities, against powers, against the rulers of the darkness of this world, against spiritual wickedness in high places" (Eph. 6:12).

The covenant couple will be first to learn this spiritual truth and fall victim to the spiritual lie, for a dark power is about to reveal its surly presence. The tragic impact will be irreversible and devastating, and consequently God's very Son must be offered up to stem the Father's wrath.

NOTES

1 Kenneth F. Weaver, "The Search For Our Ancestors," *National Geographic* (November 1985), 560-623.

2 John J. Putman, "The Search For Our Ancestors," *National Geographic* (October 1988), 439-477.

3 Ibid., 447.

4 Ann Gibbons, "Mitochondrial Eve: Wounded But Not Dead Yet," *Science* (14 August, 1992), 873.

5 Ibid., 875.

6 Ibid., 875.

7 James Shreeve, "Argument Over A Woman," *Discover* (August 1990), 52-59.

8 John E. Pfeiffer, *The Creative Explosion* (New York: Harper & Row, Publishers, 1982), 121.

9 From an exhibit in the Smithsonian Institute in Washington D.C., July 25, 1993.

10 Jacquetta Hawkes, *The Atlas of Early Man* (New York: St Martin's Press, 1976), 63.

11 Edmond Jacob, *Theology of the Old Testament* (New York: Harper & Brothers Publishers, 1958), 167.

12 Ibid., 167.

13 Dominick M'Causland, *Adam and the Adamite* (London: Richard Bentley, 1864), 302.

14 Ibid., 165.

15 C. Simon, "Stone-Age Sanctuary, Oldest Known Shrine, Discovered in Spain," *Science News*, 120. (1981), 357.

16 Dora Jane Hamblin, *The First Cities* (New York: Time-Life Books, 1973), 45.

17 Ibid., 59.

18 Ibid., 54.

19 Amihai Mazar, *Archaeology of the Land of the Bible* (New York: Doubleday, 1990), 36.

20 C. C. Lamberg-Karlovsky, and Jeremy A. Sabloff, *Ancient Civiliza-tions: The Near East and Mesoamerica* (Menlo Park: The Benjamin/ Cummings Publishing Company, Inc., 1979), 79.

21 Ibid., 99.

22 Hawkes, *The Atlas of Early Man*, 63.

23 Henry Morris, *The Genesis Record* (San Diego: Creation-Life Publishers, 1976), 84.

24 C. F. Keil, and F. Delitzsch, *Commentary On The Old Testament* (Peabody, MA: Hendrickson Publishers, 1989), 77-78.

25 Andrew Sherratt, ed., *The Cambridge Encyclopedia of Archaeology* (New York: Crown Publishers, Inc., 1980), 113.

26 George, Constable, Ed., *The Age of God Kings: Time Frame 3000-1500 BC* (Alexandria: Time-Life Books, 1987), 10.

27 S. R. Driver, *The Book of Genesis* (London: Methuen & Co. Ltd., 1938), 39.

28 E. A. Speiser, *The Anchor Bible GENESIS* (New York: Doubleday & Company, Inc., 1964), 20.

29 M'Causland, *Adam and the Adamite*, 171.

30 Driver, *The Book of Genesis*, 39.

31 "Semites" is the term archaeologists and historians use to denote not only descendants of Shem, but also descendants of Japheth, Ham, or any of Adam's line in the pre-flood period (if a person such as Adam ever existed, or there was ever an event such as the Flood). Thus, Canaanites spoke a "west semitic" language, notwithstanding Canaan was the son of Ham, according to the Bible. One might think "Hamites" would have communicated in a "hamitic" tongue. But the secular world does not recognize the Bible as being historically accurate. Therefore, "Semites" are universally recognized. "Adamites," "Hamites," and "Japhethites" are not, shall we say, "politically correct."

32 Samuel Noah Kramer, "Sumero-Akkadian Interconnections," *Genava*, n.s., 8 (1960), 272-273.

33 James B. Pritchard, *Ancient Near Eastern Texts Relating to the Old Testament* (Princeton: Princeton University Press, 1955), 60-72. Reprinted by permission of Princeton University Press. Copyright renewed.

34 Ibid., 67.

35 Ibid., 68.

36 Alexander Heidel, *The Babylonian Genesis* (Chicago: The University of Chicago Press, 1942), 56.

37 Pritchard, *Ancient Near Eastern Texts Relating to the Old Testament*, 70.

38 Ibid., 69.

39 Gleason Archer, *Encyclopedia of Bible Difficulties* (Grand Rapids: The Zondervan Corporation, 1982), 59-60.

Chapter 11

ORIGINAL SIN:
JUST SAY NO

So, this is paradise - a bountiful garden, a lovely wife, no screaming kids. Could a man ask for more? In some respects, Adam had it all. Yet he may have had an awesome responsibility he never fulfilled. As the first type of Christ, Adam may have been given a similar mission: to bring the word of God's kingdom to the polytheistic heathen living all around him. We can only guess. We can never know with certainty what it was Adam was supposed to have done, or could have done had he not yielded to Satan's odious deception so early on.

Sin Happens

Genesis 3:1-6. The wily serpent taunted and tempted Eve, who succumbed to his crafty deceit, and ate of the fruit of the "tree of the knowledge of good and evil." She, in turn, invited Adam who yielded to temptation and committed the sin of disobedience.

Those critical of a literal interpretation of Genesis pose what at first seem to be insurmountable questions, "Do you really believe a snake can talk?" and, "Aren't these passages intended to have purely symbolic meaning?"

In response, spirit beings sometimes do appear in human and animal form. For example, the Holy Spirit "descended in a bodily shape like a dove" at the baptism of Christ (Luke 3:22). In Genesis 2:26, the covenant pair was given dominion over the animals and "every creeping thing," including serpents. Eve had the power and authority to rebuke the serpent, but instead, she surrendered her God-given authority and submitted herself to Satan's influence through the guile of the serpent's words.

Had Satan appeared as an "angel of light" (II Cor. 11:14), presenting himself in splendor, a certain intimidation factor would have been introduced, his words would have seemed more credible. Eve might have been excused for believing those cunning words if Satan had appeared as a god-like being. Coming from a serpent, Eve was truly without excuse. It was simply the power of a clever argument that was her undoing.

In Luke 4:6, Satan tempted Jesus three times. Satan boasted that the power and the glory "is delivered" unto him, and that he could give it to "whomsoever" he pleased. No place in Scripture can we find where God gave him this office of power, yet he had it, and it was delegated to him. It was precisely this authority given to the covenant pair that solely through devious persuasion was delivered up on a silver platter to a lowly serpent, and it was only man's to give.

Truth or Consequences

Genesis 3:7: "And the eyes of both of them were opened, and they knew that they were naked;..."

Satan told them that if they ate of the tree their eyes would be opened. And opened they were, to see they were naked before the Lord, and short on alibis.

Genesis 3:11. God puts the question, "Hast thou eaten of the tree whereof I commanded thee that thou shouldest not eat?" Suddenly Adam needed help. He might have fallen on his face and begged forgiveness. Even after the transgression, a contrite heart might have been rewarded. Or Adam could have shouldered the blame. "Spare my wife, and I, even I, will accept responsibility for it," could have been his plea. God might have appreciated and accepted that. Even a plaintive, "The devil made us do it," would have been preferable. But there was none of that. Not repentance, or sorrow, or brokenness, but instead, panic ensued as Adam sought a scapegoat. In desperation, he turned on the only parties immediately available, his wife and his God.

Genesis 3:12: "The woman whom thou gavest to be with me, she gave me of the tree, and I did eat," Adam bleated.

The very wife Adam had been given to love and protect, he offered up to shield his guilt. And worse, Adam implicated the Giver of the gift. Squandering the opportunity to atone for his misdeed, Adam shifted the interrogation to Eve, who proved to be a quick study and countered with, "The serpent beguiled me, and I did eat" (Gen. 3:13).

And the crafty art of finger pointing entered the world.

Genesis 3:15. Speaking directly to the serpent, God announces the first prophecy of the Old Testament to the coming Messiah. "And I will put enmity between thee and the woman, and between thy seed and her seed ..."

Her "seed" was not Abel, whom Satan plotted against successfully in his bid to foil God's judgment, but Christ who through His dying would gain the victory. "It [He] shall bruise thy head," directs a death blow that would befall the evil serpent through the resurrection and final judgment after "thou shalt bruise his heel" at the cross.

> ... this clause is considered universally as referring to a Redeemer, who, in human nature, and a son of woman born, should, after partial suffering from a wicked malignant power, obtain a complete victory, and deprive it of all further means or opportunity of doing evil. The seed of the woman who was to bruise the serpent's head is connected with a singular verb and pronoun, and, denoting therefore an individual, points to Christ personally in a peculiar and emphatic sense. [1]

The pertinent point is this: biblical prophecy comes wrapped in biblical history. This is a hand in glove relationship, a muscle on bone union. Efforts to discount the historical narrative by well-intended expositors can only hurt the integrity of the Bible and cause some to question whether any parts of the Bible can be trusted.

<u>Genesis 3:16-17</u>. What were the physical results of the sin of disobedience, if any? God told Eve that He would, "multiply thy sorrow and thy conception: in sorrow thou shalt bring forth children ...," and for Adam, "cursed is the ground for thy sake ..."

The Hebrew word, *'itstsabown*, translated "sorrow," also includes pain, labor, and hardship. The operative part of the prophecy could be questioned, though, is it "sorrow," or "thou shalt bring forth children"?

Can we derive that Eve would have enjoyed painless delivery before Original Sin, or would she have had no children at all? Was the character of the ground changed due to God's curse, or were "thorns" and "thistles" outside the garden all along, which Adam had not encountered while enjoying peaceful confines?

It is entirely possible that nothing physical, biological, or horticultural happened as an immediate result of Adam's sin. It may be that the repercussions of judgment are to be understood in the spiritual sense. We cannot rule out that Adam and Eve might have suffered a biological or genetic alteration as a consequence of their sin, but neither can we assume it from the words of Scripture.

Adam must have been created biologically compatible with his neighbors outside the garden. It seems unlikely he was some kind of superman who had his genes altered after the Fall. The spiritual bond between God and man was broken. We could say then that spiritual death is surely indicated, and that physical death may or may not have been included.

If Adam had continued access to the tree of life, even after the Fall, he could have lived forever (Gen. 3:22). To prevent this, a "flaming sword" barred his entrance to the garden (Gen. 3:24). Was physical death any novelty prior to Original Sin? No, physical death has been part of the scheme of life for over 3 billion years, beginning with the most primitive of God's creations. Physical death was no stranger to early man either whose remains predate the introduction of Adam and Eve.

As a consequence of the Fall, spiritual death, a separation from God, is mandated by the text. It is evident from the immediate results that a close communal relationship between Adam and God had been severed. Possibly, physical death too may have resulted for Adam and his generations with coincident implications for mankind in general.

Death Through Sin

In Paul's letter to the Romans, he explained that due to Adam's sin, "death passed upon all men, for that all have sinned" (Rom. 5:12). Does this verse say anything about ancestry? Does our DNA contain a coding for sin? If so, could Adam be the source?

During World War II, the Nazi dictator, Adolf Hitler, was responsible for the senseless and tragic eradication of 6 million innocent Jewish civilians. Were the Jews sinners? Of course, "there is none righteous, no not one" (Rom 3:10). Was the Nazi dictator, by his insane decree, directly responsible for the death of those Jews? Yes, he was. Is it accurate to say that death passed upon all of them due to the sin of one man? Yes, it is. Were those 6 million who died descendants of the Nazi dictator? Obviously not.

In 1656, Isaac de la Peyrére argued eloquently in *Men Before Adam* that a literal interpretation of Romans 5:12-14 indicated the world was populated before Adam. The key was verse 13: "For until the law sin was in the world: but sin is not imputed when there is no law." Peyrére reasoned that the law was given to Adam shortly after his creation, and if there was "sin in the world" at that time, there must have been people to do it:

> ... it must be held that sin was in the world before Adam
> and until Adam: but that sin was not imputed before Adam;
> Therefore other men were to be allowed before Adam who
> had indeed sinn'd, but without imputation; because be-
> fore the law sins wer [sic] not imputed. 2

Although men and sin were in the world before Adam, the manner of sin was in the form of offenses against nature, violations of "natural law," and all died a natural death. It was not until God imposed moral law, with Adam the first to be subject to it, that men were capable of "legal sin," trespasses against God's law. 3

Beginning with Adam's Fall, human beings die both a natural death and a "legal" or spiritual death.

Ten years before Peyrére wrote *Men Before Adam*, the Westminster Divines penned their *Confession of Faith*. They sought to avoid any implications that all of humanity did not commence with Adam by putting the law on Moses. But if Mosaic law, and not Adamic law, was intended by Romans 5:13, it could mean that sin was not charged *before Moses*! No, the interpreters were not stepping into that trap. The Divines clearly recognized that the moral law, the "covenant of works," was given to Adam and said so:

> The rule of obedience revealed to Adam in the estate of innocence, and to all mankind in him ... was the moral law. [4]

If moral law was given to Adam, and already "sin was in the world," then wouldn't this involve people? The Westminster Divines were unwilling to entertain that possibility. They believed humanity started with Adam, and sin was passed to his posterity by "natural generation." The harmonizing device employed (although not mentioned specifically in the *Westminster Confession*) was to maintain that imputation of sin was through the law of Moses, but that it somehow applied retroactively to Adam and his descendants. This made no sense, of course, but they were torn between the illogical and the unthinkable. So, according to the Divines, the moral law was not "comprehended" until the Ten Commandments were delivered by God to Moses. [5]

Peyrére railed against the position taken by the Divines and their insistence that "the law" was the law of Moses:

> The Interpreters being between two such inconveniences, were at a stand, nor did know which way to turn themselves; But because it seemed less prejudicial to affirm, that sins were not imputed before Moses, and until Moses, than to affirm that there were any men before Adam! Therefore they preferred the first inconvenience before the second. [6]

In Peyrére's mind, since the law transgressed was the law given to Adam of Genesis, the sin was perpetrated by those who co-existed and pre-existed Adam. Sin was not imputed to those forerunners, however, until Adam disobeyed God's law.

> Before the Law of God, or till that Law of God was violated by Adam, sin and death were in the world, yet had gained no power over it: they had got no lawful possession, they had got no absolute power. The reason is, because before that time there was no Law given by God. [7]

Clearly, sin was imputed from Adam to Moses. What brought the flood? Was the flood not judgment for sin? Or, for that matter, the destruction of Sodom and Gomorrah? And if the subject of Romans 5:13 was Adamic law, the sin that "was in the world" was committed by men other than Adam.

We will never know Adam's mission on earth with certainty. It probably was intended that he was to bring news of God's kingdom to the polytheistic heathen. Adam had life to offer, perhaps tied to the tree of life some way. But regardless of what Adam was supposed to have done, however he would have done it, being human, he failed. The "second Adam" was God incarnate, and succeeded.

Adam's sin caused spiritual death upon all men, not because he was father to us all, but because he failed to be an example to us all. Adam was not the biological head of our species, but the Federal head of the human race.

The parallel Paul strikes between Adam and Christ is significant. Bloodlines are of no regard in obtaining salvation by way of the second Adam, just as bloodlines do not put us under the penalty of sin from the first Adam. Paul was the apostle to the gentiles. Indeed, he may have alluded to the gentiles being outside of Adam's line. "Nevertheless death reigned from Adam to Moses, even over them that had not sinned after the similitude of Adam's transgression" (Rom. 5:14).

It is clear from Romans 5:12-18 that all men come under the sin of Adam through his transgression, and therefore, all need a savior. What is not mandated by the text is that gentiles have Adamic ancestry, and maybe, just the opposite. It was neither by imitation nor propagation that Original Sin passed to all mankind, but by representation. Adam was the chosen intermediary; thus, we have the assurance there was no condemnation without representation.

Baking Bread

Genesis 3:19, 23: "In the sweat of thy face shalt thou eat bread ...," and, "... the Lord God sent him forth from the garden of Eden, to till the ground from whence he was taken."

Could we believe the first man on earth already knew how to use fire, construct an oven, plant and harvest grain, mill it, and prepare the flour for baking? If not, then we may conclude that Adam was not the first man in the biological sense.

Prehistoric men hunted wild game and gathered fruits and berries. Farming and raising livestock were later developments. Paleontologists have uncovered evidence that ancient peoples harvested wild wheat as far back as 9000 BC. It took a genetic crossing of goat grass and "emmer" to produce wheat bread. The earliest evidence of wheat cultivation was found in the ancient oasis of Jericho and is dated at 8000 BC. [8]

Wheat, and therefore bread, predates Adam by about 3,000 years, and nearly 4,000 years were we to use the Archbishop Ussher scale. That gives us two choices if we subscribe to a recent Adam. We can either deny the anthropological data, or allow that these agricultural developments predate Adam. If we choose the second option, Adam was surrounded at the inception, or became surrounded by people already familiar with growing grain.

Knowledge of Fire

Those who place Adam as the first human will say that Adam must have been educated by God, that he was given speech, knowledge, and mechanical skills. Certainly, God could have taught him things. However, paleontologists have uncovered evidence where early hominids manufactured and used stone tools where the use of fire was not evident.

If Adam possessed the knowledge of fire, and it was passed down through Adam's generations, how would we account for the existence of hominids smart enough to make tools, but who did not know about fire? Simply, they were forerunners, not descendants. Later, fire was discovered.

In caves near Peking, layer upon layer of ashes, baked sediments, and charred animal bones were found. These caves had been occupied by *Homo erectus*, archaeologists say, and date their findings at around 500,000 years ago. 9 If scientists are correct, then a gap of nearly half a million years would have to be conceded in the Genesis record for these cave dwellers to be Adam's descendants. If they are predecessors, however, the Genesis chronology is intact, but the knowledge of fire preceded Adam, not the other way around.

The Legend of Adapa

Cuneiform inscribed clay tablets discovered in Mesopotamian excavations have given archaeologists a picture of a region almost totally unknown only a century ago. These inscriptions have provided valuable insights into the history, religion, and racial characteristics of the people who lived there. And some of these writings contain references that seem to pertain to Adam.

The first people largely recognized as Semites, or Adamites, were the Semitic Accadians in the post flood period, though traces of these people date to possibly as early as 4000 BC. Judging from their inscriptions, the early Accadians had a triune God. From the beginning, the Accadian "trinity" consisted of El (or *Ilu*), the father god; Ea, god of the earth and creator of man; and Enlil, the god of the air. Also dating to 4000 BC, the polytheistic Sumerians were distinct from the Semitic Accadians and spoke an unrelated language.

As contact developed between these two cultures, things began to rub off. The Accadian father-god El was corrupted to "Anu" under pressure of the Sumerian "An." Enlil moved into second place, and Ea, known by the Sumerians as "Enki," dropped to third. [10]

Several fragments of the "Legend of Adapa" were taken from the Library of Ashurbanipal (668-626 BC) at Ninevah. One also was found in the Egyptian archives of Amenophis III and IV of the fourteenth century BC. [11]

According to Accadian legend, Ea created Adapa an exemplary man, endowed with "superhuman wisdom," but not eternal life. A fishing accident angered Adapa, who broke the wing of the south wind, and was summoned to heaven to appear before god Anu. Ea warned Adapa not to eat a certain food or drink any water that would be offered to him. A cautious Adapa shuns the food and water of life, whereby he would have acquired eternal life. [12]

A fragment of one record of the Adapa legend rests in the Pierpont Morgan Library. Inscribed in Amorite, a Semitic language, this is part of the translation:

> In those days, in those years, the sage, the man of Eridu,
> Ea, made him like a (riddi) among men;
> A sage, whose command no one could oppose;
> The mighty one, the Atra-hasis of the Anunaki, is he;
> Blameless, clean of hands, anointer, observer of laws.
> With the bakers, he does the baking;
> With the bakers of Eridu, he does the baking. [13]

Adam of the Bible and Adapa of Amorite legend were both human sons of God, or a god. According to the legend, Adapa was a sage, a profoundly wise man, in Eridu.

Could it be only coincidence that Adam was told "by the sweat of his face" he would eat "bread," and Adapa was a baker by trade; or that Adapa was deprived of eternal life by not eating or drinking the "food or water of life," while Adam was cut off from eating the fruit of the "tree of life"?

Regarded as a prophet or seer, Adapa had been priest of the temple of Ea at Eridu. He is described as "blameless," "clean of hands," "anointer and observer of laws." Could that also describe Adam, the first type of Christ? Also, Adam was taken from the ground; in the Hebrew: *'adam* from *'adamah*. How close phonetically is *'adamah* to Adapa?

Did Adam's Fall affect following generations? These two lines are part of one Adapa fragment:

> *[...] what ill he has brought upon mankind,*
> *[And] the disease that he brought upon the bodies of men*
> ... 14

From the Apocrypha, this Jewish tradition of the Fall is also reflected in II Esdras 7:48:

> *O Adam, what have you done?*
> *For though it was you who sinned,*
> *the fall was not yours alone,*
> *but ours also who are your descendants.* 15

Westermann concludes that in this text Adam is not understood as a "representative of mankind created by God, but as an historical individual whose 'Fall' was passed on through him to his descendants." 16

Eridu, the Home of Adapa

In 1940-41, the Iraqi government undertook the excavation of Eridu, home of Adapa.

> Here at last it was possible to trace a full and uninterrupted sequence of occupations back through the whole duration of the Al 'Ubaid period to an earliest settlement with some features so distinctive that doubts arose as to whether the name Al 'Ubaid could still appropriately be applied to it. 17

Some of the pottery found at the lowest of 19 levels of occupation was so unique that the excavators called it "Eridu ware." It was described as an "extremely fine quality monochrome-painted ware, often with a buff or cream slip." [18] There was also at the lowest level a high percentage of coarse green pottery typical of Ubaid ceramic. Enough similarities were noted between the coarse Ubaid pottery at Eridu with that of the earlier Hassuna and Samarra cultures to denote that at least some of those early settlers had been migrants from the north.

If the two different pottery styles found at the lowest level of the site are indicative of two separate cultures living side by side, one Adamite, the other Ubaid, then these pottery shards are of some importance. Quite possibly some of these remnants are from early Adamite populations.

Whatever culture was responsible for Eridu ware, Adamite or otherwise, evidently, it was supplanted by Ubaid culture as only Ubaid pottery could be found at higher levels. And just as the pottery disappeared, so perhaps did the Adamites by moving north.

Is Eridu Synonymous with Eden?

It was pointed out in the previous chapter that the Bible implies irrigation for Adam's garden, presumably via canal from Eden (Gen. 2:8, 10). In 1948-1949, Fuad Safer examined several mounds just outside of Eridu, and reported:

> The mounds were found to lie on the banks of the bed of a wide canal which, in ancient times, was undoubtedly connected with the River Euphrates. The recognition of this canal and the tracing of its course are now extremely difficult, as it has been filled with sand and soil drifted in from the surrounding plain. The course of the canal crosses the flat depression of Eridu from north-west to south-east and its nearest point to Eridu is about 3 kilometres from the south-west of that site. [19]

A branch canal from the main canal west of the city to water a garden located east of the city would have flowed through that city, exactly as stated in Genesis 2:8, 10.

The Sumerian word *edin* means "plain," "prairie," or "desert." [20] Its Accadian equivalent is *edinu*. [21] "Eden" most probably has an Accadian/Sumerian origin. Eridu is the earliest known settlement in Southern Mesopotamia at about 4800 BC. [22] The Sumerians also regarded Eridu as a sacred city. Could Eridu be synonymous with Eden? The time and place are an excellent fit.

Traveling On

Eridu is identified as the home of Adapa, but in another fragment, Adapa is called "the Erechian." [23] This, coupled with the disappearance of Eridu Ware, may indicate a relocation of the Adamites from Eridu to Erech, the Sumerian "Uruk."

> Uruk was first settled around 4200 B.C. by the Ubaid people, and at the lower levels it seems to be a characteristically Ubaid site. But beginning around 3500 B.C., there is evidence of major changes which some archaeologists believe were characteristic of a new culture and others believe represented an indigenous evolution of the "Ubaidans." [24]

Erech began around 4200 BC, some 600 years after Eridu. Again, the timing fits. The reason for moving 50 miles north could have resulted from Eridu being sacked. The kingship was overthrown; according to the Sumerian king list, a new king came to power at Badtabira. Erech was established in the pre-flood period according to Sumerian accounts, and re-established after the flood. Erech and the city of Ubaid, located only 30 miles apart, were contemporary cities situated about 140 miles southeast of Babylon. [25]

If Adam and his kin journeyed to Erech after the fall of Eridu, then this placed the children of Seth at Erech as near to the Cainite kids at Enoch (the location of the city of Enoch is discussed in chapter 12) as Brooklyn is to the Bronx. Driver took note of the remarkable similarity in names in both lines of descent. [26] Compare Sethites: Enosh, Mahalalel, Methuselah, and Lamech with Cainites: Enoch, Mehujael, Methushael, and Lamech. The similarities are understandable if they lived in close proximity.

Alias Adam

In addition to the Bible, possible variations of the name Adam appear elsewhere. On a Sumerian list of ten pre-flood kings ending in Ziusudra (the Sumerian Noah), the first king is "Alulim."

> *When the kingship was lowered from heaven*
> *the kingship was in Eridu.*
> *In Eridu Alulim became king ...* [27]

Adapa (created by the god Ea) and Alulim (king by heavenly decree) are both placed at Eridu. If Eridu is Eden, then Adapa, Alulim, and Adam could all be the same man. Conversely, if Adapa, Alulim, and Adam are the same person, and considering that the Sumerian, Accadian, and Assyrian texts place him at Eridu, then by the same token, Eridu should be Eden.

A clay tablet was recovered in excavations at Khorsabad in 1933-34. It contains a list of Assyrian kings beginning with "17 kings who lived in tents," [28] probably nomads. "Tudia" tops the list of kings followed by "Adamu," a likely namesake of his famous forefather. Farther down the list, the 38th king is "Puzar-Assur." He was one of many Assyrian kings named in honor of a more immediate forefather, Asshur of Genesis 10:11. This exact same naming pattern is consistent with a descendant of Cain in Genesis 4:22 - Tubal-cain.

Another king list is attributed to the Babylonian priest, Berossus. He recorded "Alorus" first on a list of ten pre-flood kings. According to Berossus, Alorus was "appointed by God as Shepherd of men."

The title, "the Son of God," reserved for Sumerian royalty, was used also for one called "Adamu." [29] This is a title identical to that used of Adam in Luke 3:38 where the genealogy of Christ originates with "Adam, the son of God."

In Egypt, the pyramids of kings Mer-ne-Re and Nefer-ka-Re were inscribed with a dedication dating to about 2400 BC, centuries before Abraham, and many centuries before Moses. The text speaks of a first creation and a deified "Atum" who was on a primeval hill arising "out of the waters of chaos." Among those "whom Atum begot," according to the inscription, is one named "Seth." [30]

Could Alorus, Adapa, Alulim, Adamu, Atum, and Adam be all the same person? Perhaps a better question would be, what rationale could be employed to explain away the commonalities? At least some of these secular references must pertain to the first man in biblical history. If these Egyptian, Sumerian, Accadian, Amorite, and Hebrew variations are derived from one man, the most obvious conclusion, then this not only establishes an historical Adam, a.k.a. Adamu, Atum, etc., but the time and the place is also confirmed, and in complete harmony with the Genesis text!

It should come as no surprise that Egyptian inscriptions, Sumerian legends, and Amorite epics would be based upon historical persons and events. The Sumerians probably learned about Adamic history from their own forefathers, or from their Semite neighbors who were direct descendants of Adam. Many times the Sumerians were subjects of Semite kings; the great Sargon, for example, began his reign over the entire region in 2371 BC. Adam and his successors also may have ruled over the Ubaidans who, possibly, were ancestral to the Sumerians.

The Amorites (Gen. 10:16) were descendants of Noah's grandson, Canaan. They must have passed the history of their forefathers down through their generations just as the Israelites did, but centuries of retelling took its toll. There was a special purpose in protecting the accuracy of the creation narrative handed down through the line of promise from Shem to Abraham, and through to Moses. Parallel accounts, even with embellishments and distortions, should only increase our confidence in the historical value of the Genesis narrative and affirm the astounding probability there was really a man called Adam.

NOTES

[1] Robert Jamieson, A. R. Fausset and David Brown, *A Commentary on the Old and New Testaments* (Grand Rapids: William Eerdmans Publishing Co., 1945), 57.

[2] Isaac de la Peyrére, Men Before Adam. *Or a Discourse upon the twelfth, thirteenth, and fourteenth Verses of the Fifth Chapter of the Epistle of the Apostle Paul to the Romans. By which are prov'd, That the first Men were created before Adam.* (London: 1656), 19.

[3] Ibid., 3-5.

[4] From the "Larger Catechism," *The Confession of Faith* (Philadelphia: William S. Young, 1838), 246.

[5] Ibid., 251.

[6] Peyrére, *Men Before Adam*, 19.

[7] Ibid., 19.

[8] John Wiester, *The Genesis Connection* (Nashville: Thomas Nelson Publishers, 1983), 187.

[9] John Gowlett, *Ascent to Civilization* (New York: Alfred A. Knopf, Inc., 1984), 57.

[10] Gwendolyn Leick, *A Dictionary of Ancient Near Eastern Mythology* (New York: Routledge, 1991), 37.

[11] Albert T. Clay, *A Hebrew Deluge Story in Cuneiform* (New Haven: Yale University Press, 1922), 39-41.

[12] Ibid., 40.

[13] Ibid., 41.

[14] James B. Pritchard, *Ancient Near Eastern Texts Relating to the Old Testament* (Princeton: Princeton University Press, 1955), 103.

[15] Claus Westermann, *Creation* (Philadelphia: Fortress Press, 1971), 108.

[16] Ibid., 108.

[17] Seton Lloyd, "Ur-Al 'Ubaid, Uquair and Eridu," *Iraq*, n.s., 22 (1960), 25.

[18] John Oates, "Ur and Eridu, the Prehistory," *Iraq*, n.s., 22 (1960), 33.

[19] Fuad Safer, *Sumer 6* (1950), 28.

[20] S. R. Driver, *The Book of Genesis* (London: Methuen & Co, Ltd., 1938), 38.

[21] R. Laird Harris, Gleason Archer, and Bruce Waltke, editors, *Theological Wordbook of the Old Testament*, Vol. 2 (Chicago: Moody Press, 1980), 646.

22 C. C. Lamberg-Karlovsky and Jeremy A. Sarloff, *Ancient Civilizations: The Near East and Mesoamerica* (Menlo Park: The Benjamin/Cummings Publishing Company, Inc., 1979), 110.

23 Clay, *A Hebrew Deluge Story in Cuneiform*, 41.

24 Lamberg-Karlovsky and Sarloff, *Ancient Civilizations: The Near East and Mesoamerica*, 145.

25 Yohanan Aharoni and Michael Avi-Yonah, *The MacMillan Bible Atlas* (New York: MacMillan Publishing Company, 1977), 20.

26 S. R. Driver, *The Book of Genesis* (London: Methuen & Co, Ltd., 1938), 80.

27 Thorkild Jacobsen, *The Sumerian King List* (Chicago: University of Chicago Press, 1939), 71.

28 Arno Poebel, "The Assyrian King List from Khorsabad," *Journal of Near Eastern Studies*, (1942) Vol. 1, No. 3, 252.

29 L. A. Waddell, *The Phoenician Origin of the Britons, Scots, and Anglo-Saxons* (London: Williams and Norgate, Ltd., 1924), 239, 253.

30 Pritchard, *Ancient Near Eastern Texts Relating to the Old Testament*, 3.

Chapter 12

CAIN'S LAMENT: DON'T LET THEM KILL ME!

In the previous chapter, we saw it was unlikely that Adam could have been the first farmer to grow wheat and bake bread. Probably, he learned this skill from his neighbors. It could have been a little easier to put food on the table with two strapping boys to help with the chores. But sometimes things just don't work out.

Genesis 4:2. Adam and Eve had two sons, Cain and Abel. "And Abel was a keeper of sheep, but Cain was a tiller of the ground."

Just as Adam was not the first farmer, in all likelihood, Abel was not the first shepherd. He missed that opportunity by about 3,000 years.

> At 8000 BC a few Near Eastern sites like Jericho, Mureybet, Zawi Chemi and Shanidar already appear to have had herds of sheep, goat and gazelle under conditions closely approximating domestication. 1

For Cain, seeing his brother watching sheep all day must have been galling, especially when he had to plow ground, sow seed, haul water, pick weeds, chase bugs, and reap the fields. And when that younger brother found favor from God with his offerings, while Cain found no favor as hard as he worked, well, it was more than Cain could bear.

Cain's Lament

Genesis 4:14. Cain slew Abel and lamented before the Lord, "... from Thy face I shall be hid; and I shall be a fugitive and a vagabond in the earth; and it shall come to pass, that every one that findeth me shall slay me."

Genesis 4:15. God responded to Cain's plea, "Therefore *whosoever* slayeth Cain, vengeance will be taken on him sevenfold. And the Lord set a mark upon Cain, lest any finding him should kill him."

Who was the "whosoever" to whom the Lord was referring? Cain had just eliminated his only brother. Adam and Eve have had no other children at this point. Cain could not have known whether more children would come from his parents or not. From Cain's point of view, the entire human race would have reached a dead end at that point - unless, of course, there were other human beings about.

Cain's lament proves the point. Cain's words, after the Lord banished him from Eden, were out of fear that someone would kill him. God gave him a sign. Thus we have God's confirmation that Cain's worry was valid.

Cain's fear must have been aroused at the unhappy prospect of approaching a settlement of people, probably hostile, without the Lord's protection. They would have seen Cain was a stranger, and would have tried to kill him. The Lord's action alludes to that. He certainly would not have needed any "mark" to approach his own family. So the mark of Cain must have given him some kind of identity or safe passage.

Cain's concern that a premature demise might come from human hands makes no sense, unless he was aware of other human beings, and feared them. His lament and the Lord's response indicate the co-existence of other human populations living in the same vicinity. Shields comments:

> Adam, at the time Cain slew Abel, had only these two children; consequently, Abel being slain, Adam and Cain were the only two men in the world, if Adam was the first and *only* creation of human beings. If such were the case the fear of Cain "that every one that findeth me shall slay me," must have been wholly groundless: and the reply of the Lord, that "whosoever slayeth Cain vengeance shall

be taken on him seven-fold," was at least unnecessary. Nor could there be any necessity for putting a mark upon him, "lest any finding him should kill him." For if there were no human beings then in the world but Adam, Eve, and Cain, would he not be sufficiently known to them without a mark? And would he not have been sufficiently protected from them, and their future descendants, by being driven out a "fugitive and a vagabond?" 2

Genesis 4:16: "And Cain went out from the presence of the Lord, and dwelt in the land of Nod, on the east of Eden."

Throughout the Bible the "land of Canaan" or the "land of Egypt" refers to an area populated by those particular peoples. Why have Bible interpreters not considered that the "land of Nod" might well have been populated by Nodites, who were minding their own business before Cain arrived?

In Hebrew, *nod* means "wandering." This would be an apt designation for a band of nomads who might have been in the area at the time, the word *nod* being simply a form of the word "nomad." A suitable translation would have been: ... and Cain dwelt in the land of nomads. Suitable, that is, had Bible translators considered Cain's populated surroundings.

A Wife for Cain

When Cain left the presence of the Lord and journeyed to the land of Nod - note carefully - two things happened: (1) Cain left the Lord's presence, and (2) Cain departed Eden. Next, Cain took a wife, but from where?

Some insist Cain took a sister for his wife. A niece or grand-niece would suffice, but she had to come from the covenant family to fit the presupposition that the entire human race emanated from Adam. The notion that Cain married a sibling brings a host of problems.

Did Adam and Eve have any other children at that point? Would Adam have given his daughter's hand in marriage to a man who had just slain his own brother, Adam's own son? Is it possible that Cain was married at the time he slew his brother, and departed with wife in tow? If not, could he have returned at a later time and claimed a bride from Adam's household?

These are questions that have answers, maybe not ironclad, but certainly compelling answers. If we put successive passages together, and delete some intermediate passages, it will give us a better picture.

<u>Genesis 4:1, 2, 8, 16, 17, 25, 26; 5:3, 4:</u> "And Adam knew Eve his wife; and she conceived, and bare Cain, and said, I have gotten a man from the Lord. And she again bare his brother Abel.

And Cain talked with Abel his brother: and it came to pass, when they were in the field, that Cain rose up against Abel his brother and slew him. Then Cain went out from the presence of the Lord, and dwelt in the land of Nod, on the east of Eden. And Cain knew his wife; and she conceived, and bare Enoch: and he (Cain) builded a city ...

And Adam knew his wife again; and she bare a son, and called his name Seth; for God, said she, hath appointed me another seed instead of Abel, whom Cain slew.

And Adam lived an hundred and thirty years, and begat a son ... and called his name Seth. And the days of Adam after he had begotten Seth were eight hundred years: and he begat sons and daughters."

Let us compress the data further, and list the recorded events in order of presentation:

1. Eve gave birth to Cain (Gen. 4:1).
2. Eve gave birth to Cain's brother, Abel (Gen. 4:2).
3. Cain killed Abel (Gen. 4:8).
4. Cain left Eden and settled in Nod (Gen. 4:16).
5. Cain's wife gave birth to Enoch (Gen. 4:17).
6. Cain built a city (Gen. 4:17).
7. Eve gave birth to Seth (Gen. 4:25).
8. Adam (and Eve) had other sons and daughters (Gen. 5:4).

Based upon this information, and holding assumptions to a minimum, what conclusions can we make that are within reason?

Cain was the firstborn of Adam and Eve. Again, a second time, Eve bears a child naming him Abel. Please note, Eve gave birth to *"his brother."* That phraseology tends to disallow intermediate siblings. Had daughters been born prior to Abel, then Abel would have been Abigail's brother or Edith's brother, too. In other words, the text is precisely correct if there were no other children besides the two. So, Cain was the first born, Abel, his brother, was second, and there were no other Adamite children at this point.

Some say that since the Hebrews were a patriarchal society, women were not mentioned or recorded in many instances even though roughly equal numbers of boys and girls must have been born. The argument for Cain taking a sister is based upon the assertion that Eve gave birth to daughters too, but they went unrecorded.

Let us look at that argument. Eve is most assuredly mentioned and in great detail - no patriarchal bias here. Indeed, Genesis records more quotes from her than from Adam. Following the covenant couple, we then encounter Cain and Abel in order of appearance, and then Cain's wife. She is the fifth person recorded in biblical history, yet unnamed.

Cain's line from Enoch through Lamech was set down without including females, which was the custom in recording Hebrew lineages; however, the Pentateuch might have established the custom. But, some women are named in the biblical text. Lamech took two wives, "Adah" and "Zillah" (Gen. 4:19). Zillah bore Tubal-cain and his sister, "Naamah" (Gen. 4:22).

There could be an appropriate reason why Cain's wife was not named in Genesis. She was not related to the covenant family; she had no pedigree. No sisters, wives, or women of any kind, other than Eve, are mentioned prior to Cain's departure from Eden, and taking a wife. For a sister to be available to Cain prior to his leaving Eden, we would need a Scriptural amendment.

Genesis does not record that Cain left Eden with a wife. Had that been the case, as when Lot fled the condemned city of Sodom with his wife and daughters, it surely would have warranted inclusion. If we have to make assumptions to arrive at a satisfactory answer, then let us assume the Bible is correct in its recorded facts, that what was set down is not in error, and is in the proper order. If we need additions for satisfactory answers, it would be difficult to know how much to add or where to put it.

It appears the writer of Genesis knew exactly what he (and He) was doing. We are given all the necessary and relevant facts in the correct sequence, and in sufficient, though not abundant, detail. It is entirely needless to attempt to align Scripture with human rationality by postulating an unrecorded sister as a wife for Cain.

There are sound reasons why we may conclude Cain was not married at the time he was banished, and did not take a sister or any of Adam's generations with him. First and foremost, no mention is given that Adam had any daughters until *after* Cain committed murder and departed Eden, and *after* Adam had a third son named Seth. The sheer unavailability of female relatives at that time would preclude Cain taking one for a wife.

Even though Scripture makes no mention of it, what if a sister had been available? Cain could have taken her with him to the land of Nod, could he not? No, not if Scripture is consistent. Had Cain been married at the time of his banishment, his lament before God would have made no sense. Had Cain taken a sister/wife with him out of Eden to the land of Nod, he might have registered a little concern for her welfare as well as his own.

Did Cain cry out, "Oh no, my wife and I will be killed," or, "Lord, my wife will become the world's first widow?" Why not? Because Cain was not married at the time. His concern was for himself. And if Cain's words were not enough, we have the Lord's confirming words also. Let us look back for a moment.

In Genesis 2:17, God commands Adam, "... but of the tree of the knowledge of good and evil thou shalt not eat of it."

This command was given to the man before the creation of his woman. Eve is brought into the world and presented to Adam, at which point they become married - one flesh.

The serpent tempts Eve, Eve persuades Adam, and Adam sins. In Genesis 3:14-19, the judgment is to both. It is Adam's commandment to keep, it is Adam who sins, but he and his wife, Eve, suffer the Lord's wrath. Inasmuch as they are one flesh in the sight of the Lord, God cannot punish one without punishing the other.

In Cain's situation, notice the absence of any partner with whom he could share the Lord's anger. Cain alone was judged. Had he been married, his wife also should have shared the Lord's judgment, and would have suffered a mutual pronouncement. The sin was Cain's alone, the punishment was for him alone, he raised his voice of concern for himself alone, the "mark" was for him alone, and he left Eden unmarried and alone.

A brief summary to this point: Eve gave birth to their first child, Cain. She gave birth again to a second child, Abel. Cain eliminated Abel and left Eden. Adam and his wife had another son. Seth was the "seed" God appointed instead of Abel, now deceased, or Cain, now banished.

Seth was the replacement child who came along after Cain committed murder and was banished to Nod. Had there been other children born prior to Seth, some other manner of recording his birth would have been appropriate. Also, the significance of Seth's birth would be diminished had there been other children. In a later passage, Adam had other sons and daughters, but it seems unlikely that any children, besides Cain and Abel, were born to the covenant couple prior to Seth. If Cain did not take a sister with him as he departed, his wife had to come from some other source.

One remaining possibility is that Cain departed Eden, but returned at some later date to claim his bride; however, the text precludes that too. If Cain had married one of his own relatives after being banished, it would have demanded a journey back to Eden. He would have to take a wife, and depart again. More importantly, this would have required re-entering the "Lord's presence." This could only have been done through some act of contrition or repentance. Yet no repentance of any kind is registered from Cain or any of his descendants.

The sister/wife argument also overlooks a scriptural limitation - God forbids it in no uncertain terms. (See Lev. 18:6, 9-14.) Such an incestuous union is an abomination that defiles not only the participants, but the very land (Lev. 18:24-30). Not just Cain, but Seth and at least some of Adam's married sons would have been guilty of trespassing God's law had the world been unpopulated except for Adam's clan. Even though Abraham was married to a sister, it was not by sheer necessity, and Sarah was a half-sister only.

Since we have examined the reasons why Cain must have traveled to Nod unencumbered with a spouse, we could ask who this woman was. We may never know the answer to that question completely, but we can determine who she was not. She was surely not his sister, niece, grandniece, etc.

Although Adam lived in disciplinary punishment, his generations still enjoyed the Lord's "presence." Eve makes that clear in Genesis 4:25, "For God hath appointed me another seed," named Seth. Seth fathers Enosh. "Then began men to call upon the name of the Lord" (Gen. 4:26).

Adam and his immediate family remained in the Lord's presence and protection, though Cain clearly left it, necessitating his "mark." Cain could not have come back for a wife. So from where did Cain's wife originate?

If Cain's wife did not come from Adam's line, then she either resided in a nearby settlement of people, Ubaidans probably, or she was a member of a band of nomads, whom Cain originally had feared, and for whom Cain was given his mark. The mark certainly had an effect; they not only spared Cain's life, they allowed him to marry one of their own.

Enoch City

Genesis 4:17: "And Cain knew his wife and she conceived, and bare Enoch: and he builded a city, and called the name of the city, after the name of his son, Enoch."

Perhaps partly because Cain was long lived, he was recognized as a special or unique person as evidenced by his overseeing the building of a city. A city would have been quite inappropriate for only three people, but a city might have been necessary to accommodate a growing community that included his wife's relatives.

Naming the city "Enoch" may seem like an unnecessary addendum, a bit of Bible trivia, but it is not without significance. According to the Sumerians, kingship resumed at Kish after the flood. Twenty-three kings ruled there until "Kish was smitten with weapons; its kingship to E-Anna(k) was carried." 3 In *The Makers of Civilization*, Waddell translated E-Anna(k) directly as "Enoch," reckoning it as the Sumerian equivalent for Enoch, the city built by Cain. 4

Although the flood erased early inhabitants, at least some of the pre-flood cities were rebuilt. It was at Enoch that Mes-kiag-gasher became high priest and king and reigned 324 years. 5 His son, Enmerkar, built or continued building Uruk located virtually across the street. Uruk is the biblical Erech, part of Nimrod's kingdom (Gen. 10:10). Enoch or "E-Anna(k)" (translated "the House of Heaven") is the oldest preserved temple near Uruk, and was supposedly the dwelling place of the goddess Inanna, the Accadian "Ishtar." 6

E-Anna(k), now called "Eanna" by archaeologists, has been excavated. A deep sounding was made in the Eanna precinct at Warka in 1931-32. The pottery was identified as Ubaid from level 18 up to level 14. It transitioned to the Uruk period by level 10. From Woolley's analysis, the pottery from the earliest period he found at Ur, that he called "Al 'Ubaid I," was unrepresented at Warka 7, demonstrating that both Ur and Eridu were established before E-Anna(k). And, of course, Adam's Eden would have been older than Enoch, the city Cain built.

The important point is that some of the details omitted from the biblical text are filled in by the Sumerian text, confirming not only the existence of the cities of Enoch and Erech, but also pinning down the time and the location.

Pre-Flood Cities Are Post-Flood Cities

It is especially noteworthy when we find a city such as Enoch that the Bible ties to the pre-flood period, that the Sumerians identified as existing after the flood. For one thing, it indicates the limited scope and breadth of the flood itself. Conversely, Erech, mentioned by the Bible in the post-flood period, has been excavated to reveal a culture dating to 4200 BC, over a thousand years before the flood. Likewise, Ur, the home of Abraham's youth, had pre-flood beginnings, and was contemporary with Eridu. Furthermore, Asshur built Nineveh after the flood (Gen. 10:11) on an existing city more ancient than Ur or Eridu, which dates to the pre-flood era, and had been called "Ninua" before the Semites arrived. [8]

This illustrates that at least four biblical cities dating to the pre-flood era were resettled by Sumerians and Semites after the flood. Thus we have further confirmation that the entirety of Genesis 2-11 is confined to the Mesopotamian environs, both the pre-flood and the post-flood periods; and that none of the human history contained in the Bible predates 5000 BC.

Sumerian king lists also recorded the longevity of their sovereigns. In the pre-flood period, they reigned for legendary thousands of years. After the flood, kings reigned for hundreds of years tapering off to mere mortal proportions in later periods. [9] The trend jibes with Genesis.

Although the tablets are recorded in Sumerian, some of these kings bear Semitic (Adamic) names. Cain is the only explicit pre-flood example given by the Bible, but he fits the motif of long-lived, non-Sumerian rulers who reigned over Ubaidan and Sumerian subjects. Nimrod and Asshur are biblical post-flood examples.

Adam in Time and Space

Genesis 4:21-22. Lamech, a descendant of Cain, had three sons by his two wives. Jabal "was the father of such as dwell in tents, and of such as have cattle." A second son Jubal, "was the father of all such as handle the harp and organ."

In just eight generations counting Adam, there are tents, livestock, and musical instruments; not caves, wooly mammoths, and hand axes. For many reasons, we can conclude that Adam was not

contemporary with the "Flintstones." A wealth of Stone Age artifacts have been uncovered giving silent testimony to a culture long
disappeared at this point. So where does Adam fit in the history of
man? The next verse tells us.

Genesis 4:22. One of Cain's descendants, Tubal-cain, was "an
instructor of every worker in brass and iron."

There is the proverbial smoking gun! Adam belongs after the
old Stone Ages, at the end of the Neolithic, at the threshold of a
period called the Chalcolithic when traditional stone tools were
augmented by crude copper implements. Adam's descendants saw
the dawning of the Bronze Age.

> In the initial period of the Middle Eastern civiliza-
> tions, from about 3000 BC, there was a truly remarkable
> development of metallurgy. This is seen in the beginning
> of the Bronze Age, when alloys of arsenic and copper, or
> tin and copper (in both cases known as bronze), came
> into being ... [10]

In terms of place, Southern Mesopotamia is mandated by the
Bible. Adam and his generations were surrounded by at first
Ubaidan, and then Sumerian culture:

> The fourth millennium in Sumer is one of the most re-
> markable passages in human history. Already at its be-
> ginning old settlements such as Eridu, Uruk, Ur, Lagash
> and Nippur had become substantial towns and from 3500
> BC they waxed into cities. The citizens now included
> large numbers of specialist artisans - potters, carpenters,
> makers of mudbrick, coppersmiths - and fine sculptors
> too. [11]

Sons of God

Genesis 6:1-2: "And it came to pass, when men began to multiply on the face of the earth, and daughters were born unto

them, that the sons of God saw the daughters of men that they were fair; and they took them wives of all which they chose."

The "sons of God," who are they? Some contend these are angels, perhaps fallen angels. But is that the case here? The Hebrew phrase in this passage, and elsewhere in the Old Testament, can refer to angels (Job 1:6; 2:1; 38:7; Psa. 29:1; 89:6). But the same term also describes humans who lived their lives in service to God (Deut. 14:1; 32:5; Psa. 73:15; Hosea 1:10). [12] How should it be interpreted here?

For a start, what are angels supposed to do regarding us humans? In Hebrews 1:14, "Are they not all ministering spirits, sent forth to minister for them who shall be heirs of salvation?" If that is their proper role, wouldn't it be out of character for them to be involved in these trysts? Also, even if they had the desire to sire human offspring would they be capable of that? Angels, while appearing as men at certain times, do not possess physical bodies as we do, and should not be able to father human children. [13]

Furthermore, angels do not marry. "The children of this world marry, and are given in marriage: but they which shall be accounted worthy to obtain that world and the resurrection from the dead, neither marry, nor are given in marriage; neither can they die anymore, for they are equal unto the angels and are the children of God, being the children of the resurrection" (Luke 20:34-36). And in Mark 12:25, "For when they shall rise from the dead, they neither marry, nor are given in marriage, but are as the angels which are in heaven."

Two relevant bits of information exude from these passages. Angels do not die or marry. Sons of God, who marry, should be humans. Throughout the New Testament, the term "sons of God" or "children of God" is applied exclusively to humans (Matt. 5:9; Rom. 8:14, 19; Rom. 9:26; II Cor. 6:18; Gal. 3:26). Nowhere in the New Testament do these terms apply to angels. [14]

Could they have been fallen angels? Would it have been possible that disenfranchised angels took possession of the bodies of humans in order to engage in marriage and procreation? Not likely; fallen angels or demons are not called "sons of God" anywhere in Scripture. They have forfeited that right.

Furthermore, if these had been fallen angels dabbling with the human race, then the flood would have brought only temporary relief. Demons would not drown. Any marriage-minded demons could have just waited and preyed upon the next batch of humans. Besides, the notion of demons desiring to enter into holy matrimony is a bit curious.

If the term "sons of God" refers to humans, then who could they have been? Perhaps those "who called upon the name of the Lord," the generations of Seth. 15 Then who were the "daughters of men"? The daughters of men could have been descendants from the now mixed generations of Cain, or perhaps they came from the indigenous populations that co-existed with the Adamites in the same region.

Some have contended that what has been translated "sons of God" (*bene elohim*), refers instead to sons, or servants, of pagan gods. 16 Indeed, a clear example of this can be found in Exodus 18:11 which states, "the Lord is greater than all gods (*elohim*) ..." Daughters of *ha'adam* or "the Adam," then, would be Adamite women. Using this line of logic, Genesis 6:1-2 would be translated: "And it came to pass, when the Adamites began to multiply on the face of the earth, and daughters were born unto them, that the sons [or servants] of the gods saw the daughters of the Adamites that they were fair ..."

What comes through in either translation is that there were two distinct populations, some were in the covenant line from Adam, others were not, and they were intermarrying. Lane concludes:

> ... the most obvious meaning, beyond dispute, is, that the men and women here mentioned were of *different races*, and hence that the former saw in the latter a beauty surpassing that of their own women. 17

What was the consequence of such mixed marriages? Reduced life spans.

The Years Grow Short

Genesis 6:3: "And the Lord said, My Spirit shall not always strive with man, for that he also is flesh: yet his days shall be an hundred and twenty years."

If angels or demons, who are presumed to be immortal, had been the bridegrooms, the effect should have been a prolonging not a shortening of life spans. But if these were human bonds of matrimony, then the outcome, a reduction in life spans, is what we would expect. Marriages between the long-lived descendants of Adam and the short-lived daughters from the local population, perhaps mixed with the line of Cain, produced offspring with intermediate life spans, limited eventually to no more than 120 years.

From Adam through Noah, all lived over 900 years with just two exceptions. Enoch walked with the Lord after 365 years, and Noah's father, Lamech, succumbed at 777. An early demise kept Lamech from being caught up in the deluge. Noah passed on at age 950, twenty years older than Adam when he died. Original Sin may have brought some form of death, but through ten generations, life spans were unaffected.

Noah's son, Shem, died at 600 years of age. The first downward shift in longevity came after ten consecutive generations of long life. This is an indication that Noah's wife was the probable cause, and that she had Cainite ancestry or a direct bloodline tie to the indigenous population. Although Noah was "perfect in his generations" (Gen. 6:9), no claim is made about his wife. When Noah's perfect line, through Adam and Seth, was mixed with his wife's imperfect line, their offspring had shorter life spans; and Shem, who only lived to 600, was the first to suffer a speedy demise.

Look at the succeeding generations of Shem. Arpachshad, Shelah, and Eper failed to make it to 500. The next five generations did not see a 250th birthday. Abraham and Isaac passed away at 175 and 180, respectively. Jacob died before reaching 150, and so it went. Intermarriages between the long-lived Semites and their short-lived neighbors produced children who died off at increasingly younger ages. Gradually, the results of mixing took its toll.

In light of all we now know, Genesis 6:1-3 describes intermarriages between the Adamite populations and the Ubaidans or Sumerians. From the Lagash Kinglist, Jacobsen noted that the post-flood kings of Lagash (Semites probably) not only lived extraordinarily long, they also lived extraordinarily "slowly."

> *In those days a child spent a hundred years*
> *In diapers (lit. "in <bits> of the wash")*
> *After he had grown up he spent a hundred years*
> *Without being given any task (to perform)*
> *He was small, he was dull witted*
> *His mother watched over him.* [18]

In contrast to the long-lived post-flood kings, skeletal remains at the pre-Adamic city of Catal Huyuk yielded an average age at death of about 34 years old. [19] Has archeological discovery confirmed the mixing of covenant generations with non-covenant generations?

> Another break in cultural tradition and an acceleration in civic advance began around 4000 BC. Some historians believe that these changes were due to the arrival of the Sumerians on the plain, perhaps again coming from the north. Others do not accept a distinct immigrant group but see the Sumerians as an amalgam of all the prehistoric peoples of the region. The language, however, when it came to be recorded, does suggest a Sumerian tongue overlaying a more primitive one that might well have been that of the Ubaidans. It also contains some Semitic elements and it is likely that Semites were already drifting into the valley from the north. [20]

Technically, "Semites" refers to the descendants of Shem because universally historians do not recognize Adam or Noah. Is it possible, though, that the Sumerian language contained not "Semitic elements," but Adamic or pre-flood Accadian language elements? If so, then the presence of loan words in the Sumerian language supports Genesis 6:1-3. Adamites were mixing with Non-Adamites.

We do know that after the flood, Semites spread out and encountered peculiar populations in their path (Gen. 15:20, Deut. 2:10, 11, and Josh. 13:12, for example), but 4000 BC is pre-flood history.

Noah's Wife the Key to the Origins Question

There is a key to how all of us human beings could be related to a primordial ancestor (or a small group of ancestors) who lived 100,000 years or more, and yet Adam, father to the Semites among others, was specially created according to the Bible. The key lies with Noah - or rather Noah's wife.

Noah was five hundred years old when he begat Shem, Ham, and Japheth (Gen. 5:32). We have no way of knowing how old Noah's wife would have been, but she could have been in her teens at the birth of Shem. The flood took place in Noah's 600th year, giving him three grown sons to help build the ark.

Noah's wife was still alive after the flood (Gen. 8:16, 18), although there were no more children. If short-lived, she would have been past her childbearing years when the flood ended. This is the last passage about her. We do not know when she died, but Noah's drunkenness and lying naked in his tent (Gen. 9:21) might have resulted partly from his despondence after her death.

It is possible that Noah's wife might have died before reaching her 120th birthday due to mixed parentage, or Non-Adamite parents, although she could have lived a little longer. Cain, who may have been a distant ancestor, might have lived into his 900's. But as Cain's sons (and daughters) took short-lived wives (and husbands) with ancient ancestry, succeeding generations died off at younger ages. A kind of dilution effect occurred. Or Noah's wife may have had no Cainite ancestry, in which case long life would have been unlikely.

Notice that there are significant differences between Noah's family life and that of the preceding patriarchs. Starting with Adam, every one of the first nine patriarchs was less than 200 years old when he became a father, [21] whereas Noah did not have children until he was 500 years old! This is too great a difference to be without significance.

Then also, all of Noah's predecessors had sons and daughters. Noah had no other children after the three boys, Shem, Ham, and Japheth. Even the patriarchs for seven generations after Noah, "begat sons and daughters" (Gen. 11:11-25). The indication is that Noah's wife was past her childbearing years after the flood causing Noah to be unique in parenthood compared to both his forefathers and descendants. The most obvious reason would be that Noah's wife was unique, she was short-lived, and from outside the covenant line.

History Lessons for Shem

There is another reason why Noah's wife may have come from a mixed line including Cain. Had Cain been banished from Eden and disappeared altogether, then Chapter 4 of Genesis could not have been written without supernatural dictation. Who held the knowledge of the history of Cain's offspring recorded in Genesis 4:17-24? Noah's wife, a possible descendant, is a likely contributor for the narrative of Cain and his generations. An alternate candidate would be Shem's wife.

Shem is the probable source of the knowledge of pre-flood events passed down through the Semites. Shem learned the chronology back to Seth and Adam (in Genesis 5) from his father, Noah, or his grandfather, Lamech, or his great grandfather, Methuselah. Observe that the names and ages are given in a matter-of-fact manner and are bereft of any detail. Take Genesis 5:6-12 as an example. Note that the pattern is repeated in each succeeding generation. No details are provided with the exception of Enoch, and the narrative includes no dialogue.

In Genesis 5:6-9, "And Seth lived an hundred and five years, and begat Enosh: And Seth lived after he begat Enosh eight hundred and seven years and begat sons and daughters; and all the days of Seth were nine hundred and twelve years; and he died. And Enosh lived ninety years ...," and so on.

Contrast this repetitive, even tedious, litany of Seth's generations with the richly detailed account of Cain's generations found in Genesis 4.

"And the Lord said unto Cain, Where is Abel, thy brother? And he said, I know not: Am I my brothers keeper? And He said, What hast thou done? The voice of thy brother's blood crieth unto Me from the ground" (Gen. 4:9-10).

"And Lamech said unto his wives, Adah and Zillah, Hear my voice; ye wives of Lamech, hearken unto my speech: for I have slain a man to my wounding, and a young man to my hurt. If Cain shall be avenged sevenfold, truly Lamech seventy and sevenfold" (Gen. 4:23-24).

The most probable reason the accounts of the generations of Seth and the generations of Cain vary in both format and content, would be simply because they came from two different sources. It is likely the account in Genesis 5 came from Shem's father, grandfather, or great grandfather from the line of Seth. The history of Cain probably came from Shem's mother (unnamed), who could have been a Cainite. Again, Shem's wife is an alternative source. Lest there continue to be skeptics, ponder the words of Moses in this next verse.

Giants in the Earth

Genesis 6:4: "There were giants [*Nephilim* in the Hebrew] in the earth in those days; and also after that, when the sons of God came in unto the daughters of men, and they bare children to them, the same became mighty men which were of old, men of renown."

The term *Nephilim* means little more to us today than does "the land of Nod" or "gopher wood." These are words of antiquity and will always remain obscure. And yet, the text tells us of some kind of men who were different, were of ancient origin, and were well known at the time. It is true the biblical narrative is brief in the extreme on the subject of these "children" and the *Nephilim*, but whoever the Nephilim were, "fallen ones," "apostates," perhaps, they do not appear to be from Adam's race. [22]

So why isn't this all spelled out in the Bible? Van Amringe confronted this problem of apparent omission:

> If the creation of Adam and Eve was a remedial measure,
> by the All-wise Creator, to remedy the vices of the people
> then in being; -- and Adam and Eve were made in the
> image and likeness of God, for this purpose; and the more
> important purpose of furnishing a proper line of beings,
> through whom God himself was, at a future time, to be
> born upon the earth, as the greatest of all remedial mea-
> sures, -- we can see that it was not important to disclose to
> Moses any more of the creation of that period than what
> immediately related to Adam, as the progenitor of the Sav-
> ior. [23]

Some have interpreted Genesis 6:4 to mean that "gods" or angelic beings, derived from Genesis 6:1-3, married human beings
that gave birth to a race of giants. [24] The text simply does not
allow that, especially in light of New Testament clarifications. Genesis 6:4 implies that the Nephilim were already on the earth when
the fore-mentioned marriages took place, and they "were of old,"
indicating their ancestors pre-existed this period of time. If we
interpret these passages literally, then the existence of Pre-Adamites
living in close proximity is not merely a possibility, but an ines-
capable conclusion.

Adam was specially created, responsible to God, and yet biologically compatible with other humans who were already living
in the region at the time of Adam's introduction. Adam could not
possibly have started all the Near East peoples, let alone the human race, due to his late entry, but rather he was placed in a locale
which was already populated by that time.

Cain entered the world of flesh and took a wife. Sons from
Seth's line, including perhaps, male descendants from other sons
and daughters of Adam, took wives from one or more of the local
farming communities and possibly from the mixed line from Cain.
Conversely perhaps, Adamite women were chosen for wives by
Non-Adamite men, or maybe, some from column A and some from
column B. This caused subsequent generations to be mixed, being
both of "spirit" and of "flesh." Van Amringe summed it up:

Thus, then, the fear of Cain, "that every one that findeth me shall slay me;" -- his marriage in the land of Nod, before Adam and Eve had daughters; -- the men, and giants, of those days, as distinguished from the "Sons of God," and the wickedness which prevailed among them, -- all appear to point to a race of human beings, prior to the creation of Adam and Eve. [25]

Judgment of the Flood

Genesis 6:12-13: "And God looked upon the earth, and, behold, it was corrupt; for all flesh had corrupted his way upon the earth. And God said unto Noah, the end of all flesh is come before Me; for the earth is filled with violence through *them*; and behold, I will destroy *them* with the earth."

The word translated "them" is *ha'adam*, literally "the Adam," or "the Adamites." Those from Adam were the target of the flood which terminated a multitude of men and all of Adam's descendants, except for Noah and his family. The judgment of the flood was brought down upon the Adamites, those who were accountable and capable of sin. Other unfortunates in the vicinity were also swept away.

Although Noah was a direct descendant of Adam, and "perfect in his generations" (Gen. 6:9), we are not told from where his wife or his son's wives originated. Someone had to be the source of the narrative of Cain and his line. The most probable source is Noah's wife, or maybe, the wife of Shem. Noah's wife, and the wives of his sons, must have had mixed Cainite ancestry, or simply came from the local populace.

Adamic ancestry accrues to only a small percentage of people scattered around the globe today. Traces of Adam's genes might be found in present-day Arabs, Jews, and their offshoots, and should have been present in early populations such as Amorites, Hittites, Canaanites, and others. But even among modern peoples who might have Adamic blood ties, there is still no escaping ancient history, and with it, ancient ancestry.

Some may claim Adam as a forefather, others may doubt it, and most just don't know. But, because of the intermarriages, even those who feel they can boast of biblical ancestors can presume they have roots that reach back 100,000 years, even 4 million years, and perhaps, beyond.

NOTES

1 Jacquetta Hawkes, *The Atlas of Early Man* (New York: St. Martin's Press, 1976), 46.

2 W. F. Van Amringe, *An Outline of a New History of Man* (New York: Baker & Scribner, 1848), p. 53.

3 Thorkild Jacobsen, *The Sumerian King List* (Chicago: The University of Chicago Press, 1939), 85.

4 L. A. Waddell, *The Makers of Civilization* (New Delhi: S. Chand, 1968), 62.

5 Jacobsen, *The Sumerian King List*, 85.

6 Samuel Noah Kramer, *From the Poetry of Sumer* (Berkeley: University of California Press, 1979), 174.

7 Lloyd, "Ur-Al 'Ubaid, Uquair and Eridu," *Iraq*, n.s., 22 (1960), 24.

8 I. E. S. Edwards, C. J. Gadd and N. G. L. Hammond, eds., *The Cambridge Ancient History* Vol. I, Part 2, (Cambridge: Cambridge University Press, 1971), 730.

9 Jacobsen, *The Sumerian King List*, 77-91.

10 John Gowlett, *Ascent to Civilization* (New York: Alfred A. Knopf, Inc., 1984), 180.

11 Hawkes, *The Atlas Of Early Man*, 64.

12 Gleason Archer, *Encyclopedia of Bible Difficulties* (Grand Rapids: Zondervan Publishing House, 1982), 79.

13 Ibid., 79.

14 Ibid., 80.

15 Ibid., 80.

16 Edward William Lane, Reginald Stuart Poole, ed., *The Genesis of the Earth and of Man* (London: Williams and Norgate, 1860), 75.

17 Ibid., 77.

18 Thorkild Jacobsen, "The Eridu Genesis," *Journal of Biblical Literature* 100/4 (1981) 520-521.

19 Dora Jane Hamblin, *The First Cities* (New York: St Martins Press, 1973), 58.

20 Hawkes, *The Atlas of Early Man*, 63-64.

21 Two patriarchs exceed 200 years in the Septuagint. Adam was 230 years old at the birth of Shem who in turn was 205 years old at the birth of his son, Enosh.

22 Lane, *The Genesis of the Earth and of Man*, 80-81.

23 W. F. Van Amringe, *An Outline of a New Natural History of Man* (New York: Baker & Scribner, 1848), 61.

24 Ronald S. Hendel, Chapter 13 "When the Sons of God Cavorted with the Daughters of Men," *Understanding the Dead Sea Scrolls*, ed., Hershel Shanks (New York: Random House, 1992), 167-177.

25 Van Amringe, *An Outline of a New Natural History of Man*, 57-58.

Chapter 13

THE GENESIS FLOOD: GLOBAL OR LOCAL?

A few years ago, a couple of limbs blew down from an enormous pecan tree that grew in my yard. I bought a chain saw, cut the limbs into firewood-sized pieces, and stacked them in a storage shed. As I labored for three days, it gave me only a tiny glimpse of what it must have been like to build an enormous ark.

Noah had no chain saw. Unlike my logs, which needed to fit a fireplace, the beams for the ark had to be cut with enough precision that it could carry an immense load and withstand the pressures of a lengthy flood. Coating with tar would have done little good if the timber had not been hewn with care. We can assume Noah had assistance, but regardless of the amount of help, constructing a huge water-tight vessel would have been virtually impossible without metal saws, axes, hammers, and such.

The necessity of semi-modern tools to accomplish such a feat of construction places a limit as to how far back into history the flood could have taken place. The deluge had to have happened in relatively recent times when copper or bronze was in use.

It matters little whether the flood was of short duration, or whether it was a protracted year long odyssey. The task for which the boat was constructed requires an ability to produce it, which puts the flood event somewhere into fairly recent history, if we can call around 5,000 years ago "recent." Since modern man was already racially divided and had covered the globe sparsely by this late date, the flood must have been narrowly confined.

In *A Survey of Old Testament Introduction*, Archer condenses Ramm's conclusions concerning the inherent weaknesses in the global flood argument. In Archer's words, "Formidable scientific problems are raised by a universal flood according to Ramm's summary:" [1]

(1) According to the best estimates, to cover the highest Himalayas would require eight times more water than our planet now possesses.

(2) The withdrawal of so great a quantity of water constitutes an almost insuperable problem, for there would be no place to which it could drain off. The mechanics of this abatement of water would certainly be difficult, for the atmosphere could not possibly hold that much water in evaporated form, and it is doubtful if any underground cavities in the earth could receive more than a small fraction of this additional volume of water.

(3) Scarcely any plant life could have survived submersion under salt water for over a year, and the mingling of ocean water with the rain must have resulted in a lethal saline concentration, even though the mixture would have been considerably diluted. Practically all marine life would have perished, except those comparatively few organisms which can withstand tremendous pressure, for 90 percent of present marine life is found in the first fifty fathoms, and many of these species cannot survive distant migration from their native feeding grounds. Presumably the fresh water fish would have died, even though the salinity might have been high enough to support salt-water fish.

(4) Certain areas of the earth's surface show definite evidence of no submersion. For example, in Auvergne, France, there are reportedly cones of loose scoria and ashes from volcanoes thousands of years older than the flood, and yet they show no signs of having been washed or disturbed by flood waters. [2]

A World Flood?

Strickling tackled the problem of Noah's flood, and compiled sixty-one legends of flooding catastrophes from all over the world, and found interesting similarities as well as striking differences. A favored family saved in a boat has a basis in mythology from various parts of the world. A remnant population of an unspecified number, using other means of survival, also has a basis in mythology. Through statistical techniques, he concluded:

> Either catastrophic flooding of global or near-global dimensions occurred more than once, or there were more survivors of the Great Deluge than one crew, or both. [3]

Strickling reasoned that a one-time universal event with a family of eight as sole survivors was not feasible. If Noah's flood was a universal event, there were numerous survivors in many locales; or perhaps, flooding occurred many times during man's history, and survivors used various means of escape, or both.

Taking the counter argument, Montgomery observed:

> The destruction of well nigh the whole human race, in an early age of the world's history, by a great deluge, appears to have so impressed the minds of the few survivors, and seems to have been handed down to their children, in consequence, with such terror-struck impressiveness, that their remote descendants of the present day have not even yet forgotten it. It appears in almost every mythology, and lives in the most distant countries, and among the most barbarous tribes. [4]

Montgomery included a schematic summary taken from Byron C. Nelson's *Deluge Story in Stone* that plots out the existing mythological accounts on a graph showing both similarities and discrepancies. Montgomery, who had access to similar data as Strickling, reached the opposite conclusion. He endorsed a universal, one-time only event with eight survivors versus Strickling's conclusion that such could not have been the case. So, what is amiss?

Flood Stories

What about the flood stories that permeate the mythology of remote populations? Interestingly, the differences more than off-set the similarities. Nelson's schematic of 41 flood myths shows that just nine of them mention saving animals. However tempting it might be to attribute all those ancient stories to a one-time global catastrophe to conform with the traditional interpretation of the Genesis flood, a literal reading of Genesis does not require it, and the unyielding revelations of nature and history disavow it.

It should not surprise us that floods punctuate the distant past of many present-day civilizations. A look at a map of the United States, paying particular attention to its cities, shows that early European settlers located their population centers usually on rivers or at river junctions. Concerns for drinking water, bathing, washing clothes, irrigation, and transportation overpowered concerns about flooding.

Why should primitive men think differently? It would have been only natural for early tribes to camp along rivers, and to be swept away upon occasion. Indeed, besides tribal warfare, what other kinds of catastrophes could there have been in ancient days? It is to be expected that survivors would be most vocal in recounting a devastating flood to following generations. The *Interpreter's Dictionary of the Bible* deflates the idea that flood stories from different parts of the world might be related to the biblical account.

> At one time this widespread distribution of a flood tradition was considered proof of the historicity of the biblical account, which with some expected modification had spread throughout the world as people migrated from their original homeland in the Near East. This notion has necessarily been given up. We know, e.g., that numerous

peoples have no flood legends in their literature. Flood stories are almost entirely lacking in Africa, occur only occasionally in Europe, and are absent in many parts of Asia. They are widespread in America, Australia, and the islands of the Pacific. In addition, many of the known flood legends differ radically from the biblical story and stand independently of it and of one another. Many do not know a world-wide flood at all, but only a local inundation.... Often the heroes save themselves in boats or by scaling mountains, without intervention by the gods. Further, only a few of the flood stories give the wickedness of man as the cause for the Flood.... The duration of the Flood, if given, varies from a few days to many years. Facts of this kind disprove the claim that the biblical account is the parent of all flood stories. [5]

Also, we need to consider the impact early missionaries had on the mythology of primitive peoples. The biblical account of the great flood, related by missionaries, may have become interwoven with ancient tribal stories to produce hybrid myths that would parallel the Genesis narrative more closely. According to Gaster no flood story can be traced in Sanskrit until after elements of the Aryan civilization began to arrive in India. The Nestorian Christian missionary attempts in China stand out as the source of the flood story among the Lolos people. [6] Archer admits:

The list of descendants in the respective lines of Ham, Shem, and Japheth as recorded in Genesis 10 does not permit any easy identification with the remoter races who lived in the lower reaches of Africa, Far East Asia, Australia, and the Americas. Particularly in the case of Australia, with its peculiar fauna indicating a long period of separation from the Eurasian continent, the difficulty of assigning either the humans or the subhuman population with the passengers in the ark has been felt to be acute. [7]

In other words, the Bible is silent on any possible relationship between the descendants of Noah and the Black Africans, or the Mongoloid race, or the native Americans who descended from the Asiatics, or the Aborigines who populated Australia, or even the blond-haired Scandinavians, not to exclude any racial group. [8] That squares exactly with what we know about the antiquity of those races of peoples who were far distant from the Mesopotamian valley by 5,000 years ago. From C. S. Coon:

> Since the beginning of agriculture no new subspecies (of man) have arisen; the principal changes that have taken place have been vast increases in the numbers of some populations and decreases to the threshold of extinction in others. All this points to one conclusion: the living subspecies of man are ancient. The origins of races of subspecific rank go back into geological antiquity, and at least one of them is as old by definition, as our species. [9]

The Issue of Race

A caution flag should be raised at this juncture lest anyone make unwarranted racist conclusions. Adam's niche in history is tens of thousands of years after the advent of modern *Homo sapiens* and the branching out of the great races. With no ancestral ties, Adam could not have been from any particular race. No one can say Adam was Caucasian, for example, even though present-day peoples with possible blood ties to Noah's three sons have Caucasoid features.

The wives of Noah and the wives of his three sons are the key. Considering the likelihood these women had blood ties to the distant past, then this almost assuredly mandates Caucasian ancestry for them. At that point in history, the resident populations in the Mesopotamian valley, the Sumerians, were dark-haired, light-skinned Caucasians. [10] And, possibly through intermarriage, this is the same racial type of modern-day Semites.

Although the human genome mapping project is still incomplete at this writing, researchers have produced a detailed physical map of the human Y chromosome. [11] When fathers pass the X chromosome they have daughters; the Y produces sons. It is now possible to construct male family trees using the Y chromosome as has been done using mtDNA leading back to ancient "Eve." Soon mitochondrial Eve may have a companion in "Y chromosome Adam." Hopefully, they both will date to the same time period so they will have a chance to know each other a little.

It is problematic whether one could verify Adamic ancestry by analyzing the Y chromosome trees of Jews, Arabs, Armenians, and others, who would be obvious candidates as descendants of Adam. The reason is that Noah's male descendants were few in number compared to the large numbers of surrounding indigenous populations. Each time one of Noah's male descendants went childless or had only daughters, that line became a Y chromosome dead end.

We know the Sumerians and Semites became a mixed population early on. The Sumerians even acquired the sacrificial system, offering lambs and unblemished oxen and goats to their deities. [12] But regardless of the ultimate origins of these two ethnically distinct peoples, we may find that no present-day males can be traced to Noah and Adam. Obviously, if we found men today whose Y chromosome trees extended back to only 7,000 years or so, then the explanation in this book would become instant fact. Still, discovering any males with such a marker seems rather unlikely.

Nature's Evidence

The island of Madagascar, to cite one example, with its populations of lemurs found no place else on earth, [13] puts a damper on any notion of a massive worldwide flood after the advent of hominids. Madagascar drifted away from the mainland of Africa about 165 million years ago, even before monkeys and apes had come into existence. [14]

Today, Madagascar is inhabited by 28 species and 40 subspecies of lemurs that are totally unique to that island. [15] The present day lemur populations, dramatically different from other animal populations found elsewhere in the world, denies the possibility of a global flood with the termination of all animal populations during the last 100 million years.

A survey of other island populations, each with its own unique animal life, weighs against any global catastrophe taking place during the time of human history. The existence of kiwis in New Zealand, kangaroos and koala bears native to the continent of Australia, to point out just a few examples, precludes a global destruction after the advent of hominids whenever and however they arrived, not to mention Noah who was a veritable "Johnny-come-lately."

The following excerpt is taken from the *Encyclopedia Britannica*:

> Each Yucca moth species is adapted to a particular species of yucca (plant). The moths emerge when the yucca flowers open. The female gathers pollen from one flower, rolls it into a ball, flies to another flower, lays four or five eggs, and inserts the pollen mass in the opening thus formed. The larvae eat about half the 200 seeds produced by the plant. The yucca can be fertilized by no other insect, and the moth can utilize no other plant. [16]

The idea of a Yucca moth hopscotching the globe with its yucca plant partner in tow did not sit well with Strickling, who made the following comment:

> Given a universal destruction by the Flood, the relationship must have come into existence afterward; it would be absurd to claim that the two partners migrated in unison from the Ark in its Old World resting place to their home in the New, surviving only in the latter. The alternative is something less than universal destruction. [17]

The Flood in Perspective

How does the notion of "something less" than a global flood square with the Genesis account? Halley addressed that issue:

> "All the high mountains that were under the whole heavens, were covered. And all flesh died that moved upon the earth" (Gen 7:19, 21). This, doubtless, is the very language in which Shem related, or wrote, the story of the Flood to his children and grandchildren. He told it as he saw it. Are we to interpret his language according to his own geography, or present day geography? The whole race, except Noah and his family, were destroyed. To destroy the race it was necessary for the Flood to cover only so much of the earth as was inhabited. Accepting the Bible account as it is, there had been only TEN generations from Adam, the first man. How could ONE family, in TEN generations, with primitive modes of travel, populate the whole earth? Most likely the race had not spread far outside the Euphrates basin. [18]

Halley does not seem to be aware of extra-Noahic populations, but he does opt for a non-global flood. The following comes from Archer:

> In explanation of this assertion (that the flood was not necessarily universal) it needs to be pointed out that the Hebrew 'eres, translated consistently as 'earth' in our English Bibles, is also the word for 'land' (e.g. the land of Israel, the land of Egypt). There is another term, tebel, which means the whole expanse of the earth, or the earth as a whole. Nowhere does tebel occur in this account, but only 'eres, in all the statements which sound quite universal in the English Bible (e.g., Gen. 7:4, 10, 17, 18, 19). Thus, Genesis 6:17c can be rendered: 'Everything that is in the land shall die' - that is, in whatever geographical region is involved in the context and situation. [19]

To reiterate: an unenlightened Bible translation has made victims of us all. The word "earth," synonymous with "globe" or "planet," is a permissible translation of the Hebrew word 'erets, from Genesis 1:1 to 2:4, even though this last verse is transitional, and shifts focus to the immediate area where Adam was created, where the flood took place, and where the tower of Babel was built.

From Genesis 2:5 to 12, words such as "land," "region" or "territory" fit the context better than the word "earth," with the possible exception of Genesis 8:22 and 9:13. Cain was not driven off "the face of the earth" (Gen. 4:14), just out of the vicinity of Eden. Clouds never cover the globe completely (Gen. 9:14), only a segment of land. The planet was not divided in Peleg's days (Gen. 10:25), simply the immediate region.

Undoubtedly, the Old Testament writers had no concept of the earth as a round globe with a circumference of 25,000 miles. What we can visualize as the earth today is entirely different from what they could have pictured as a definition of the word. Could the Hebrews or Egyptians or any other Near Eastern cultures have envisioned the world then as we know it exists today, with polar ice caps and oceans covering three-fourths of the surface, massive land continents, and numerous oceanic islands burgeoning with unique faunal populations?

The notion of a global flood, based solely on the Genesis narrative, fails on two counts: (1) the word translated "earth" in Genesis can mean "land," and (2) any word which might have defined "earth" would not mean then what it means today.

Revelations in Clay

When the British first began excavations in 1849 in what used to be called Assyria, diggers filled crates with some 25,000 clay tablets they shipped off to the British Museum, thinking they were just decorated pottery. They exercised little care loading them into baskets with the inevitable result:

... the voyage was more disastrous for those documents
than had been the taking of Ninevah by the Medes. [20]

What ended up in London in a pile of dust were the broken
fragments of the most valuable history of Mesopotamia. This had
been the library of King Ashurbanipal who collected meticulously
and stored his treasure trove in 668-626 BC. It was the Assyrian
king himself who decided, or helped decide, which historical documents were important enough to copy for posterity. Candidate literature included writings from before the flood. We might have
found them fascinating, but the king was unimpressed:

> I study stone inscriptions from before the flood, which
> are obtuse, obscure and confused. [21]

Considering pre-flood inscriptions to be of little value, the king
did not allot precious time and material to making copies. What
was preserved and recovered are somewhat ragged legends pieced
together with words inserted sometimes where they look like they
belong. Still, the preponderance of material collected over the years
is more than sufficient to document a flood episode closely paral-
leling the Genesis account.

Ziusudra

Kingship was "lowered from heaven" and established at Eridu.
The Sumerian king list continues:

> *In Eridu Alulim became king*
> *and reigned 28,800 years.* [22]

Obviously, the length of rule is suspect. The reign of all the
pre-flood kings, recorded in Sumerian measure, runs into the thousands of years. Using a sexagesimal system, the years recorded
for the ten kings ending with Ziusudra are in multiples of 60 or 60
squared. [23] Probably there is something we do not understand about

their recording of years, but suffice it to say the Sumerians believed these kings ruled for long periods, and therefore, must have lived many years.

In succeeding verses, the kingship was transferred, through warfare most likely, from Eridu to Badtabira, Larak, and Sippar, ending in Shuruppak with the reign of Ubartutu, the eighth king. Suruppak was the son of Ubartutu, and Suruppak's son was Ziusudra. 24 There is some disagreement in lists discovered. Some name eight kings, some list ten, some lists end with Ziusudra. (This is discussed further in chapter 15.) In Jacobsen's *Sumerian King List*, this narrative follows the list of pre-flood kings:

> *The Flood swept thereover,*
> *After the Flood had swept thereover,*
> *When the kingship was lowered from heaven*
> *The kingship was in Kish.* 25

A tablet recovered from Nippur contained about 300 lines with the first 37 missing. Following is part of the flood account originally written in Sumerian cuneiform:

> *The gods of heaven and earth [called upon] the names of*
>   *Anu and Enlil.*
> *Then did Ziusudra, the king ... build a mighty ...*
> *Obeying in humility and reverence, [he] ...*
> *... the gods, a wall ...*
> *Ziusudra, beside it, stood and hearkened.*
> *'Stand on my left by the wall ...*
> *By the wall will I speak a word to thee, [hearken to*
>   *my speech]*
> *[Give] ear to my commandment:*
> *By our ... a flood [shall invade] the places of worship,*
> *To destroy the seed of mankind ...*
> *This is the decision, the decree of the assembly [of the*
>   *gods].*
> *By the command of Anu (and) of Enlil ...*
> *Their kingship, their dominion [shall be abolished].'*
>   (Break of about forty lines.)

The hurricanes, in monstrous fury, attacked as one;
At the same time the deluge swept over the places of
worship.
Then, for seven days (and) seven nights,
The flood was poured out over the land,
(And) the great ship was tossed by the hurricanes
upon the mighty waters.
Utu came forth, he who sheds light over heaven and earth.
Ziusudra opened a window in the great ship;
Utu, the hero, cast his beams into the interior of the giant
boat.
Ziusudra, the king, fell on his face before Utu.
The king kills an ox, slaughters a sheep. [26]

The Sumerian version concludes with eternal life being granted to Ziusudra from on high. As the "preserver of the seed of mankind," he was given a place to dwell. The end of Ziusudra's reign at Shuruppak concludes the Pre-dynastic Period in Mesopotamian history. The Early Dynastic Period (E.D. I) began at Kish after the flood. The start of E.D. I is dated at 2900 BC, and this date is confirmed by the flood layer found at Shuruppak during archaeological excavations, dated independently at 2900 BC.

Atrahasis

The Atrahasis version has been pieced together from Babylonian and Assyrian recensions. Out of an original of approximately 1245 words inscribed, only a scant 170 remain. Atrahasis means "Exceeding Wise" and is a title used also for Adapa. [27] Thus a link exists between Atrahasis, who survived the flood, and Adapa, who missed out on eternal life, that mirrors the biblical link between Adam and Noah. (See Genesis 1:27-28 and Genesis 9:9-10.)

Following is a portion of Atrahasis:

The land became wide, the peop[le became nu]merous,
The land bellowed like wild oxen.
The god was disturbed by their uproar.

[Enlil] heard their clamor
(And) said to the great gods:
'Oppressive has become the clamor of mankind.
By their uproar they prevent sleep.'
            (Some lines are skipped here.)
'Wall hearken to me,
Reed hut, guard well all my words!
Destroy the house, build a ship,
Renounce (worldly) goods,
Keep the soul alive!
The ship thou shalt build.
            (The following is condensed.)
That [ship] shall be an ark, and its name
        Shall be 'Preserver of life.'
[...] ceil (it) with a mighty cover.
[Into the ship which] thou shalt make,
[Thou shalt take] the beasts of the field,
        The fowl of the heavens.
Atra[hasis] opened his mouth to speak,
[Say]ing to Ea [his] Lord:
"I have never built a ship [...]
Draw a design [of it on the gr]ound.
That, seeing the [des] ign, I may [build] the ship. [28]

The remainder of the account speaks of drought and pestilence that falls upon the land and lasts for years. The flood ensues finally, destroying those not seeking refuge on the ship. [29]

Utnapishtim

After the flood, kingship was restored at Kish. When Kish was "smitten with weapons," the monarchy moved around a bit, and kings were installed in other cities.

Legends of kings and heroes of ancient times were popular stories worth repeating. One such was Gilgamesh, and tablets of the Gilgamesh epic have been found all over the region. The legendary Gilgamesh, fifth post-flood king of Uruk (the biblical

Erech), [30] was a folk hero to the Sumerians in much the same manner as the mythical "King Arthur" is to the English. Folk telling obviously inflated Gilgamesh's deeds and travels; nevertheless, he is frequently referenced, leading historians to believe he was an historical personality as well.

Inscribed in Accadian, a semitic language predecessor to Hebrew, this story tells how Gilgamesh was grief stricken at the death of his good friend Enkidu. This caused him to reflect upon his own mortality, and to realize that everyone's days were numbered; well, almost everyone. There lived a man in recluse who had survived a devastating flood, was reputed to have lived an exceedingly long life, and was even thought to possess eternal life - a gift from the gods. He was called Utnapishtim, literally "Long-lived."

Dalley in her book *Myths from Mesopotamia* goes further:

> ... it is just possible that an abbreviation of (Uta)-na'ish(tim) was pronounced 'Noah' in Palestine from very early times. [31]

The eleventh tablet of the Gilgamesh epic contains the encounter of the renowned Gilgamesh with the legendary Utnapishtim. Following is a condensation:

> *Gilgamesh said to him to Utnapishtim the Faraway:*
> *"As I look upon thee, Utnapishtim,*
> *Thy features are not strange at all; even as I art thou*
> *My heart had regarded thee as resolved to do battle*
> *[Yet] thou liest indolent upon thy back!*
> *[Tell me,] how joinst thou the Assembly of the gods,*
> *In thy quest of life?"*
> *Utnapishtim said to him, to Gilgamesh:*
> *"I will reveal to thee, Gilgamesh, a hidden matter*
> *And a secret of the gods will I tell thee:*
> *Shuruppak-a city which thou knowest,*
> *(And) which on Euphrates' [banks] is situate-*
> *That city was ancient, (as were) the gods within it,*
> *When their heart led the great gods to produce the flood.*
> *[There] were Anu, their father,*

Valiant Enlil, their counselor,
Ninurta, their assistant,
Ennuge, their irrigator.
Ninigiku-Ea was also present with them;
Their words he repeats to the reed-hut:
'Reed-hut, reed-hut! Wall, reflect!
Man of Shuruppak, son of Ubar-tutu,
Tear down this house, build a ship!
Give up possessions, seek thou life.
Forswear (worldly) goods and keep the soul alive!
Aboard the ship that thou shalt build,
Her dimensions shall be to measure.

(Skipping and condensing a little.)

The little ones [carr]ied bitumen,
While the grown ones brought [all else] that was needful.
One (whole) acre was her floor space,
Ten dozen cubits the height of each of her walls,
I laid out the contours (and) joined her together.
I provided her with six decks,
Six 'sar' (measures) of bitumen I poured into the furnace,
Three sar of asphalt [I also] poured inside.
Whatever I had of all the living beings I [laded] upon her.
All my family and kin I made go aboard the ship.
The beasts of the field, the wild creatures of the field,...
He who orders unease at night, showers down
        a rain of blight.'
I watched the appearance of the weather.
The weather was awesome to behold.
I boarded the ship and battened up the entrance.
To batten down the (whole) ship, to Puzur-Amurri,
        the boatman,
I handed over the structure together with its contents.
With the first glow of dawn,
A black cloud rose up from the horizon.
Inside it Adad thunders,
While Shullat and Hanish go in front,
Moving as heralds over hill and plain.
Erragal tears out the posts;
Forth comes Ninurta and causes the dikes to follow.
For one day the south-storm [blew],

Gathering speed as it blew, [submerging the mountains],
Overtaking the people like a battle.
The gods were frightened by the deluge,
And shrinking back, they ascended to the heaven of
        Anu.
Six days and [six] nights
Blows the flood wind, as the south-storm sweeps the land.
When the seventh day arrived,
The flood (-carrying) south-storm subsided in the battle, ...
I looked about for coast lines in the expanse of the sea:
In each of fourteen (regions)
There emerged a region (-mountain).
On Mount Nisir the ship came to a halt.
Mount Nisir held the ship fast,
When the seventh day arrived,
I sent forth and set free a dove.
The dove went forth, but came back;
Since no resting-place for it was visible, she turned round.
Then I sent forth and set free a swallow.
The swallow went forth, but came back;
Since no resting-place for it was visible, she turned round.
Then I sent forth and set free a raven.
The raven went forth and, seeing that the waters had
        diminished,
He eats, circles, caws, and turns not round.
Then I let out (all) to the four winds
        And offered a sacrifice.
I poured out a libation on the top of the mountain." 32

Without question, similarities stand out between this account and the Genesis record. To point out two that may not be quite so obvious, note that Noah was found lying in his tent in a drunken state (Gen. 9:21-23), and Utnapishtim was in similar repose. "Noah walked with God" (Gen. 6:9), and Gilgamesh inquires how Utnapishtim attained "the assembly of the gods."

Xisouthros

Berossus, a Babylonian priest in the third century BC, compiled a history he titled in Greek, *Babyloniaka*. [33] Although no copies have survived, the Jewish historian, Josephus, and other Greek writers have referred to it or included quotations. In the Berossus version, Xisouthros was directed in a dream to set down a pre-flood history, bury it, and build a boat. He was to stock the boat with animals, relatives, and friends, and ride out the impending flood. The boat landed in Armenia, whereupon birds were released to test the terrain.

Xisouthros, along with his wife, daughter and boatman, disappear into immortality while the survivors travel to Babylon to rebuild the destroyed city. This account written in Greek is thought to have derived from the legend of Ziusudra written in ancient Sumerian.

Comparison and Parallels

It cannot be ignored that the extra-biblical versions parallel the biblical version to varying degrees. Details differ, but a common thread can be seen that suggests a common source. God is (or the gods are) displeased with the state of humanity. A man and his family are singled out. That man is warned of an impending flood, builds a boat and loads it with animals and birds. They ride out the storm, coming to rest in a hilly or mountainous region. Birds are released and a sacrifice or libation is offered. In the end, God (or the gods) "smell the sweet savor."

Differences are also noticeable. When I was a college student taking American History, the Civil War was considered an important event worthy of study. My class was assigned a number of books by several authors, and our exam consisted of comparing the different versions.

Even though the Civil War had begun less than a hundred years earlier, the rationale for it, the importance of the various events, the political climate, etc., all varied widely according to each author's philosophical point of view. It cannot be denied that the Civil War took place, but when seen through different eyes, the accounts were dissimilar. Likewise, attempts to write off these flood stories as erroneous mythology, or merely pagan lore, are unjustified.

For one thing, we have the flood layers themselves. Many of the cities named in stories about the flood have been excavated to reveal the actual clay layers between remnants of ancient populations. Furthermore, the layers at Kish, Shuruppak, Uruk, Lagash, and the higher layer at Ur, all date to roughly the same period, 2900 BC. From the evidence we can infer that all of the flood stories, both biblical and extra-biblical, were predicated on an event.

The event, a flood, was talked about and written about, and the accounts were passed down through many generations. Whether Gilgamesh ever encountered Utnapishtim is as problematic as Godzilla meeting King Kong. Who knows? In all likelihood an imaginative scribe concocted it. But what is conspicuous is that he drew upon established traditions. Elements of the story were in circulation.

The biblical narrative was predicated on the same event, but corresponding accounts were passed along separate channels. The history of Noah's flood comes to us thanks to Moses, we believe, who used source materials at hand. Moses, a discerning servant of God, was the filter through which any polytheism was screened out.

Many historians believe the Hebrew version in the Bible was derived from pagan mythology. This belief is unfounded. What should be seen is that the Mosaic account of the flood, as well as the epic myths, are all based upon a like event in history, a sort of "shared common ancestry" as it were.

After comparing the Babylonian epic with the biblical account, Wiseman concluded:

> Any similarities with the Genesis record have to be over-laying extraneous matter which forms the bulk of the poem; such can best be explained as due to both versions going back to a common primary fact. [34]

And this is the point precisely. The ultimate source of all the accounts is the event itself, a massive flood, that impacted the entire region so heavily it remained a staple both of folk lore and of Genesis.

NOTES

1 Bernard Ramm, *The Christian View of Science and Scripture* (Grand Rapids: Eerdmanns, 1954), 244-46.

2 Gleason Archer, *A Survey of Old Testament Introduction* (Chicago: Moody Press, 1974), 210-211.

3 James E. Strickling Jr., *Origins - Today's Science, Tomorrow's Myth* (New York: Vantage Press, 1986), 33-39.

4 John Warwick Montgomery, *The Quest for Noah's Ark* (Minneapolis: Bethany Fellowship, Inc., 1972), 23.

5 *Interpreter's Dictionary of the Bible*, Vol. II (New York: Abingdon Press, 1962), 280.

6 Theodore H. Gaster, *Myth, Legend and Custom in the Old Testament* (New York: Harper & Row, 1969), 96, 355, sec. 38, n. 6.

7 Archer, *A Survey of Old Testament Introduction*, 213.

8 Jacquetta Hawkes, *The Atlas of Early Man* (New York: St. Martins's Press, 1976), 54-55.

9 C. S. Coon, *The Origin of Races* (New York: Alfred A. Knopf, Inc., 1962), 20.

10 George Constable, ed., *The Age of God-Kings: TimeFrame 3000-1500 BC* (Alexandria: Time-Life Books, 1987), 10.

11 Leslie Roberts, "Two Chromosomes Down, 22 to Go," *Science* (22 October 1992), 28-30.

12 Clarence Elwood Keiser, *Selected Temple Documents of the Ur Dynasty* (New York: AMS Press, 1983), 52, and George A. Barton, *The Royal Inscriptions of Sumer and Akkad* (New Haven: Yale University Press, 1929), 223.

13 Allison Jolly, "Madagascar's Lemurs On the Edge of Survival," *National Geographic* (August 1988), 132-161.

14 Ibid., 140.

15 Ibid., 141.

16 From "Incurvariidae" in *Encyclopaedia Britannica*, 15th edition (1984), V: 326.

17 Strickling, *Origins - Today's Science, Tomorrow's Myth*, 39.

18 Henry H. Halley, *Halley's Bible Handbook* (Grand Rapids: Zondervan Publishing House, 1965), 74.

19 Archer, *A Survey of Old Testament Introduction*, 210.

20 Andre Parrot, *The Flood and Noah's Ark* (New York: Philosophical Library, 1953), 22.

21 Ibid., 13.

22 Thorkild Jacobsen, *The Sumerian King List* (Chicago: The University of Chicago Press, 1939), 71.

23 Lloyd Bailey, *Noah: The Person and the Story in History and Tradition* (Columbia: University of South Carolina Press, 1989), 123.

24 Parrot, *The Flood and Noah's Ark*, 42.

25 Jacobsen, *The Sumerian King List*, 77.

26 Parrot, *The Flood and Noah's Ark*, 35-37.

27 James B. Pritchard, *Ancient Near Eastern Texts Relating to the Old Testament* (Princeton: Princeton University Press, 1955), 104.

28 Ibid., 104-105.

29 Edmond Sollberger, *The Babylonian Legend of the Flood* (London: The Trustees of the British Museum, 1962), 26.

30 C. Leonard Woolley, *The Sumerians* (New York: AMS Press, 1929), 22.

31 Stephanie Dalley, *Myths from Mesopotamia* (New York: Oxford University Press, 1989), 2.

32 Pritchard, *Ancient Near Eastern Texts Relating to the Old Testament*, 93-97.

33 Bailey, *Noah: The Person and the Story in History and Tradition*, 13.

34 D. J. Wiseman, *Illustrations from Biblical Archaeology* (London: Tyndale Press, 1958), 8.

Chapter 14

NOAH AND FAMILY: A VOYAGE TO REMEMBER

In 1928-29, Leonard Woolley excavated the ancient city of Ur in Southern Mesopotamia. It had once been a thriving Sumerian port city situated on the Persian Gulf. The build-up of silt over centuries moved the coastline south, and now the long-abandoned ancient ruins lie many miles inland. In the hot desert sun and swirling dust, Wooley's grimy work crews stabbed their spades time and again, digging deeper and deeper, filling their buckets with loose desert sand.

We Have Found the Flood

"The graves of the kings of Ur," Woolley called them, yielded their treasures of precious trinkets one by one. Cups and goblets, vases and jugs, lyres and harps, gold, bronze, and silver pieces of adornment were excised from rooms lined by walls of stone. Grisly skeletons were wrested from their dark resting places and brought to searing sunlight.

In the summer of 1929, toward the end of their season, Woolley's native digging crews made one last probe, a little deeper than the last plundered grave, to see what was in store for the following year. They found more pieces of artifacts, much to their pleasure, beneath the foundations of the lowest tomb.

Encouraged by these finds, Woolley next wanted to know how far down they had to go before the treasure trove would end. Shafts were sunk carefully. Sand and debris were brought up for close examination. Woolley had dated the lowest tomb to 2800 BC. Now, inch by inch and bucket by bucket he was traveling back in time.

Clay tablets started appearing among the loose debris. The inscriptions bore characters deemed to be even older than his previous finds. He had reached 3000 BC by his reckoning, and with appetites whetted afresh, still there was more to come. Shafts were sent down deeper still.

At last the floor was reached, the journey had ended, they stood at the bottom, or so they thought. Then Woolley noticed the soil had changed from sand to clay. Upon close examination, Woolley determined that the clay had once been dissolved. What was water-laid clay doing in the middle of a desert beneath these tombs?

His first thoughts were that the clay layer must have been set down when the Euphrates river overflowed its banks sometime long before civilization had begun. Yet, the elevation seemed too high for that. His next step was to measure the depth of the layer of clay. To his amazement, nearly 10 feet of clay was discovered before reaching another level of civilization beneath the clay layer.

Once again artifacts were brought to the surface for scrutiny. Another discovery was made; the bits and pieces of pottery were uneven, a sign that they were made by human hands alone, unaided by the potter's wheel. These painted potsherds were from a civilization even more primitive than the ones he had already uncovered, and later were identified as "Ubaid." [1]

Woolley reasoned that two dissimilar civilizations separated by 10 feet of water-laid clay could mean only one thing. An ecstatic Woolley reached a conclusion and sent a telegram that electrified the world of 1929, though to a lesser degree than did the stock market crash of that same year. "We have found the Flood," announced an ebullient Woolley.

Elated by his find, Wooley encouraged excavators at other sites to look for flood layers. Sure enough, flood layers or more cautiously "sterile stratum" of various thicknesses were found. At last, thought Woolley, archeology had established firm evidence for what had long been a controversial Bible story - the great flood. But that euphoric feeling was not to last.

Dating archeological digs in the absence of deposits of volcanic ash lacks the kind of precision archaeologists prefer, but nevertheless, the thick flood stratum Woolley found at Ur was placed at the early fourth millennium, about 3800 BC. Notwithstanding, a higher flood level also was uncovered dated to about 2700 BC, but it had been discounted as too little and too late. 2

Langdon and Watelin excavated Kish in 1928-29. They dated the bottom layer which amounted to about one foot in thickness to 3300 BC. This seemed to lend support to Woolley's claim, even though the dates were 500 years apart. The thickest layer at Kish was at a higher level, however, and assigned a similar date to the thinner layer found at Ur.

Mallowan, who excavated the more northern city of Ninevah, uncovered several strata of mud and riverine sand totaling six feet in depth. Diplomatically, he called this not a flood, but a "pluvial interval," and placed it at the fourth millennium, similarly dated to Woolley's layer. But then, flood deposits at Kish, Shuruppak, Uruk, and Lagash were considered and a consensus put all of these layers at nearly a thousand years later than Woolley's renowned find, averaging around 2900 BC. 3

Archaeological Flood Dates

PERIODS	DATES	Ur	Kish	Shuruppak	Uruk	Lagash	Nineveh
Early Dynastic	2470 / 2800		▬▬				
Jamdat Nasr	2800 / 3000			▬▬▬	▬▬	▥▥▥	
Uruk	IVth Millennium						
Obeid		▬▬▬					▬▬
Halaf							

Comparative chronology of Mesopotamian sites which contain flood deposits. From André Parrot, The Flood and Noah's Ark, *p. 52. Courtesy of SCM Press.*

This prompted a debate turning on who had uncovered the biblical flood - the most important flood in human history - and it left Woolley in a bit of a quandary. After all, he had wasted a lot of time and energy if his trumpeted flood deposit was from the wrong flood.

As it turned out, the flood layer Woolley thought was from Noah's flood was dated far too early in relation to the other sites, while the higher layer he had discounted, that was dated closer to the flood layers from the other sites, seemed puny by comparison. Ironically, the lower, earlier, and thicker layer Woolley thought was from *the flood* resulted apparently from merely a flood. And conversely, the higher, later, and thinner layer he thought was from only a flood, may have been left by *the flood*. Such is the life of an archaeologist who goes public a smidgen too early.

Sorting Out the Floods

The importance of the dating and the distribution of the flood deposits at the six sites cannot be overlooked. Ninevah is located farthest north, and according to the Bible, was not settled by Semites until Asshur led his expedition in the post-flood era (Gen. 10:11).

At the time of 3800 BC, Ur was still a Ubaidan city on the Persian Gulf, which became a Sumerian city later on. The local floods at these two sites, Ur and Ninevah, would have had devastating effects on the local populations, but neither city is likely to have experienced any Adamite migrations at those early dates.

The cities that bore the brunt of a massive flood all at the same time, around 2900 BC, were the middle cities of Southern Mesopotamia which would have contained Adamite populations. What may seem odd at first is that no flood layer was found in the excavation at Eridu. Eridu may have experienced flooding, or it may have been spared.

First of all, Eridu was the southern-most city, located on the Persian Gulf at that period, roughly 20 miles from the Euphrates. So, Eridu may have remained high and dry. Also, Adamite populations may have migrated north by then, settling at Erech and elsewhere as indicated by the Adapa fragment (discussed in chapter

11). As retribution for the sins of the Adamites, the flood may have involved no more territory than that containing Adamite populations. Probably, Ninevah, Ur, and maybe even Eridu, were outside the boundaries of Adamite settlements at the time of the flood.

Area of Flooding

Map of Mesopotamia, showing places connected with major flooding. From André Parrot, The Flood and Noah's Ark, p. 33. Courtesy of SCM Press.

Of equal importance to the finding and dating of the flood deposits connected with Kish, Shuruppak, Uruk, Lagash, and possibly Ur, is the total absence of flood layers found anywhere else in the Near East. What has been hard for many to accept is that the flood of Noah's day was entirely local to Southern Mesopotamia. According to Bright:

> A number of sites in Mesopotamia, of equal or greater antiquity, have been excavated down to virgin soil, and no evidence of flooding came to light at them. Perhaps the most important of these is Eridu, located only seven miles away from Ur. Equally serious is the fact that no site in Syria or Palestine, where archaeologists were equally active during the early part of the present century, has yielded a "flood layer." In these two countries some of the oldest towns in the world have been excavated ...(and) show no evidence of a flood ... [4]

Parallel Accounts Confirm the Genesis Flood

The present-day distribution of animals around the globe, along with the fossilized remains of their early ancestors in the same locales, precludes a global flood at such a late date in earth history. Furthermore, the Genesis flood harmonizes with the local legends. By comparing the Genesis flood narrative with its counterparts we will see just how closely they follow suit.

Hard Hearts

Genesis 6:5: "And God saw that the wickedness of man was great in the earth, and that every imagination of the thoughts of his heart was only evil continually."

Utnapishtim: "*Shuruppak - a city which thou knowest, (and) which on Euphrates [banks] is situate - that city was ancient, (as were) the gods within it, when their heart led the great gods to produce the flood.*" [5]

Although the Atrahasis account lists "clamor," "uproar," and maybe, overpopulation as the reason for bringing the flood, the Gilgamesh tablet agrees with Genesis. It was hardened "hearts" that brought on the flood.

Man Destroyed

Genesis 6:7: "And the Lord said, I will destroy man whom I have created from the face of the earth;..."

Ziusudra: "*By our ... a flood [will sweep] over the cult centers; to destroy the seed of mankind ...*" 6

Those who occupied the land are to be destroyed from the land. Although the Sumerians were first to set down their version of the flood, they also were first to know how to write. Probably they learned the flood story from the Semites, although they must have had some firsthand knowledge. Sumer flourished centuries before the flood.

A Favored Servant

Genesis 6:8: "But Noah found grace in the eyes of the Lord."

Atrahasis: "*[Ea] opened his mouth, [say]ing to his servant: Thou sayest 'let me seek ...' The task which I am about to tell thee guard thou well: Wall hearken to me, reed hut, guard well all my words! Destroy this house, build a ship, renounce worldly goods, keep the soul alive.*" 7

The subterfuge, apparently, was that Ea was in the inner counsel of the gods, and was not supposed to reveal the decision to bring the flood and destroy man to any mere mortal. Ea speaks to the wall so that Atrahasis may overhear the words, and escape death.

Pitching the Ark

Genesis 6:14: "Make thee an ark of gopher wood; rooms shalt thou make in the ark, and shalt pitch it within and without with pitch."

Utnapishtim: *"I laid out the contours (and) joined her together. I provided her with six decks, dividing her (thus) into seven parts. Her floor plan I divided into nine parts. I hammered water-plugs in her. I saw to the punting-poles and laid in supplies. Six 'sar' (measures) of bitumen I poured into the furnace, three sar of asphalt [I also] poured inside."* 8

Pitching boats on the inside was unusual from what we know about early ship building. Yet both accounts confirm this was done.

Designing the Ship

Genesis 6:15: "And this is the fashion which thou shalt make it of: The length of the ark shall be three hundred cubits, the breadth of it fifty cubits, and the height of it thirty cubits."

Atrahasis: *"[Say]ing to Ea [his] Lord: I have never built a ship [...]. Draw a design [of it on the gr]ound that, seeing the [de]sign, I may [build] the ship."* 9

Utnapishtim: *"The ship that thou shalt build, her dimensions shall be to measure. Equal shall be her width and her length. (Skipping some lines.) Ten dozen cubits the height of each of her walls, ten dozen cubits each edge of the square deck."* 10

The dimensions differ, as do the proportions. But remember, the Sumerians used a sexagesimal system that affects measures of time and measures of length. What is important is that both accounts describe a huge ship requiring a lot of work to build.

Saving Family and Animals

Genesis 6:18-20: "But with thee will I establish my covenant; and thou shalt come into the ark, thou, and thy sons, and thy wife, and thy sons' wives with thee. And of every living thing of all flesh, two of every sort shalt thou bring into the ark, to keep them alive with thee; they shall be male and female. Of fowls after their kind, and of cattle after their kind, of every creeping thing of the earth after his kind, two of every sort shall come unto thee, to keep them alive."

<u>Atrahasis</u>: *"[Into the ship which] thou shalt make, thou shalt take the beasts of the field, the fowl of the heavens.* (Skipping a few lines.) *Aboard her [bring] thy grain, thy possessions, thy goods, thy wife, thy family, thy relations, and thy craftsmen. Beasts of the field, creatures of the field, as many as eat herbs, I will send to thee and they shall guard thy door."* 11

The epic accounts appear to save more than eight people unless Noah's family, "thy relations," were also "craftsmen." In the eleventh tablet of Gilgamesh, the boatman plays a part, aiding Gilgamesh in finding the plant of eternal life. Still, nothing precludes Noah's sons from being boatmen or craftsmen. And certainly, we have no reason to regard the epic accounts as "true" even if based upon an actual occurrence.

Fountains of the Deep

<u>Genesis 7:10, 11, 12</u>: "And it came to pass after seven days, that the waters of the flood were upon the earth. In the six hundredth year of Noah's life, in the second month, the seventeenth day of the month, the same day were all the fountains of the great deep broken up, and the windows of heaven were opened. And the rain was upon the earth forty days and forty nights."

<u>Utnapishtim</u>: *"A black cloud came up from out the horizon. Adad thunders within it, while Shullat and Hanish go before, coming as heralds over hill and plain; Erragal tears out the masts, Ninurta comes along (and) causes the dikes to give way; ..."* (Skipping some lines.) *"Six days and six nights the wind blew, the downpour, the tempest, (and) the flo[od] overwhelmed the land. When the seventh day arrived, the tempest, the flood, which had fought like an army subsided in (its) onslaught."* 12

The phrase "fountains of the deep" (Gen. 7:11; 8:2) has been a major contributor to the global flood concept. Visions of great, oceanic, water-spewing volcanoes have been conjured up to rationalize this phrase, and to account for the massive amount of water needed for a universal deluge.

Analyses of the flood layers at the excavated city sites found only those elements that could be expected from the waters of the Euphrates. No remains of any salt water creatures were present which indicates none of the floods involved sea water. 13

In chapter 11, we examined the Septuagint version where the word "fountain" appears rather than "mist" in Genesis 2:6. We saw this referred to an irrigation system in all likelihood. Here "fountains of the deep" again points to irrigation. The Hebrew word for "deep" can mean the sea, it can refer to subterranean waters, or it can mean the depths of a river. In the Atrahasis epic, the phrases "fountains of the deep" or "fountain of the deep" appear four times. In all instances, fountain(s) pertain to "fields," as in this example:

> Be[low] the fountain of the deep was stopped, [that the
>   flood rose not at the source].
> The field diminished [its fertility]. 14

From the consistency in usage, we can see these were canals or levies used for irrigation. In the Gilgamesh account, Ninurta was the "lord of the wells and irrigation works." 15 So, we now know precisely what the phrase "fountains of the deep" means. The expression is defined by usage, and was employed by Semites long before Moses used it in the flood narrative. It was the overflowing rivers that caused the dams, dikes, and irrigation canals to burst open, flooding the land. We can now properly interpret "fountains of the deep" as a reference to irrigation, which clearly mandates a local flood.

Duration of the Flood

One of the differences between Genesis and the epic accounts is the length of time that it rained upon the earth, forty days versus seven days. In chapter 8 of this book, the Hebrew use of perfect or prophetic numbers was addressed. Here are two "perfect" examples; 7 and 40 are just good numbers. Genesis 1-11 is full of them. All the accounts, both inspired and uninspired, use these perfect, prophetic, or magical numbers.

It has been suggested that the god Ea might correspond to "Yahweh" in Hebrew, but it's also possible that Ea was a primitive form of another name found in Matthew 1:23 - "Emmanuel." The Accadians sometimes denoted Ea by his sacred number, "forty." [16] The number forty has special significance in the life of Christ. He spent forty days fasting and being tempted in the wilderness (Mark 1:13), and His ascension into heaven was forty days after the resurrection (Acts 1:3).

Did the original Hebrew text use the number forty in Genesis in a similar sense that a subsequent scribe took to mean the number of days? Alas, we do not know. Suffice it to say that it rained upon the earth for a perfectly long enough period of time to accomplish God's plan.

Scope of the Flood

Genesis 7:19, 20: "And the waters prevailed exceedingly upon the earth; and all the high hills, that were under the whole heaven, were covered. Fifteen cubits upward did the waters prevail; and the mountains were covered."

Some Genesis commentators have seized on these passages to assert that the high mountains of the whole earth were covered to a depth of fifteen cubits (about 22 feet). Where the water would have come from is problematical, as well as what became of it.

Measuring any depth at all raises a question. How would the passengers from inside the ark have any idea what the depths were? The Gilgamesh epic speaks of "punting" and a boatman. Long poles would afford a means of measurement. Some means of directing the boat could have been helpful as it may have traveled against the flow. Since the slope of the land is from north to south, had the ark been like a floating log, it should have been swept downstream, and ended up in or near the Persian Gulf. There had to have been enough water to get the ark to hill country where it came to rest, and some rudimentary means of navigation may have been employed, though conceivably, a raging south wind might have done the work.

Again, the word for "mountains" and "hills" is the same in Hebrew. If the flooding was restricted to the region of the Mesopotamian valley, then the "mountains" submerged by the flood could have been the lower mountains of the region surrounding the valley, or it may signify the lower foothills at the beginning of a mountain range.

As to the language used to describe the flood, it would make no difference whether the flood, in fact, was global or local. From the standpoint of the passengers on the ark, the description is entirely true and accurate in either case. These verses do not oblige us to ponder whether the Rockies, or the Andes, or the Urals, or the Himalayas were under water.

Considering that mountains were not inundated by the flood, as the evidence indicates, in no way should that impugn the accuracy or inerrancy of Scripture. From Noah's and Shem's viewpoint, the text describes their situation and surroundings in terms we might have used had we been passengers on the ark ourselves.

For example, say we heard an emotional outpouring from someone who had just fled from a burning building. If that person exclaimed, "There was fire and smoke everywhere," would anyone rebuke him for speaking inaccurately? Who would chide a shaken survivor with, "Now, you don't mean 'everywhere,' do you? You meant only inside the building." In this hypothetical situation, who would not know instantly what was intended by the word, "everywhere"? We make interpretations from context every day. Are we to be any less sensible when Scripture is the case in point?

Why should Scripture, yes "inspired" Scripture, be interpreted differently? Humbly, obediently, reverently, judiciously, and studiously - yes, but we need not abandon our common sense. A regional flood confined to the Mesopotamian valley fits all the requirements of accuracy and inerrancy that anyone should expect.

Death of Man

Genesis 7:21, 22: "And all flesh died that moved upon the earth, both of fowl, and of cattle, and of beast, and of every creeping thing that creepeth upon the earth, and every man: All in whose nostrils was the breath of life, of all that was in the dry land, died."

Utnapishtim: *"And all of mankind had returned to clay."* 17

In the Gilgamesh narrative, the goddess Ishtar "cried out like a woman in travail, 'Like the spawn of fishes they fill the sea!' " 18 "The Anunnaki gods ('his followers,' 19 angels perhaps) weep with her. The gods, all humbled, sit and weep."

Although rampant polytheism may have been a chief reason for the judgment of the flood, Van Amringe credited "the children of Adam intermarrying with the daughters of men," 20 as the fatal sin; by reason of which "all flesh had corrupted his way upon the earth" (Gen. 6:12).

> There were then residents of Asia, probably near or about the Euphrates; consequently, it was not necessary that the punishment of the Deluge should be more extensive than the prevalence of the wicked beings who had become corrupt. If, therefore, there were other men in the world besides Adam and his descendants, -- and if the Deluge did not prevail over all parts of the Earth at the same time, -- it follows, that, although all the descendants of Adam, except Noah and family, were destroyed, there may have been others, in other parts of the earth, who escaped. 21

The only reason under the sun for considering the flood to be a global catastrophe, obliterating all the world's humanity and all the world's air-breathing land animals, is the biblical narrative itself. One cannot help but get the impression that the flood encompassed more than just the Mesopotamian valley. But the last phrase, "of all that was in the dry land, died" should help us keep our perspective. Mesopotamia, present-day Iraq, is a desert, and a desert is a "dry land."

Long Flood or Short Flood?

As to how long the occupants of the ark had to endure their adventure, the epic narratives offer no time whatsoever between the seven day flood and the boat coming to rest in a mountainous region. Genesis is more explicit.

Genesis 7:24: "And the waters prevailed upon the earth an hundred and fifty days."

The great flood of 1993 that inundated parts of Missouri, Iowa, Illinois and adjacent areas in the Midwest can give us a measure of how a local flood can last over an extended period of time. Some fields and farms were underwater for weeks. Yet Noah was afloat for five months according to the Genesis text. Could a local or regional flood last that long?

It has been a long-held presumption that the rain must have fallen nonstop for the first forty days and nights causing the flood, which then stayed on the earth for an additional 110 days, for a total of 150. An alternate view is that out of a total of 150 days there were forty days of rain, which could have been stretched out over two or more floods. A sequence of floods may have required Noah and his family to stay aboard the ark until all the flooding ceased.

Yet, this is an answer only necessary for those having a mind set like that of most Greek, German, and English-speaking peoples. We are precision thinkers, and our language reflects it. We can take 150 days, and with the help of a pocket calculator, we can dissect that time period into hours, minutes, seconds, or even nanoseconds, if we like. And we like that kind of thing. But if we have a problem with exactness, it is our problem. It is not a Hebrew problem.

In the mind set of the ancient Hebrews, there is a beauty in perfect or prophetic numbers that overcompasses precision. In chapter 7, we looked at this verse, "These six things doth the Lord hate: yea, seven are an abomination unto him" (Prov. 7:16). We like to know whether it is six or seven. To the Hebrews, seven includes six.

In Genesis 8:13, the "waters were dried up from off the face of the earth" on the first day of the first month. Noah looked, and "behold, the face of the ground was dry." In the next verse, the earth was dried on the 27th day of the second month.

The face of the earth was dry; then 57 days later, it was dry too. This not a problem in ancient Hebrew; it is people like us who make it a problem. We will never know whether there was a series of floods, or if standing water covered the entire region continually for five months, or whether 150 is simply the sum of perfect numbers, two 40's and ten 7's.

The Ship Comes to Rest

Genesis 8:4, 5: "And the ark rested in the seventh month on the seventeenth day of the month, upon the mountains of Ararat. And the waters decreased continually until the tenth month: in the tenth month, on the first day of the month were the tops of the mountains seen."

Utnapishtim: "*Upon Mount Nisir, the ship grounded. Mount Nisir held the ship that it moved not.*" 22

As a further step toward reconciliation, let us dispel the myth that the ark came to rest high on the 17,000 foot Mount Ararat. The Genesis text, using the plural "mountains" (or hills), identifies no particular mountain, but points toward Armenia, "Ararat" being identical with the Assyrian "Urartu" which broadly embraces that region. 23 Mount Nisir from the Gilgamesh epic is also recorded in the annals of King Ashurnasirpal II of Assyria. This is a low-lying mountain at the beginning of the Zagros range situated south of where the Little Zab joins the Tigris, near the 9,000 foot Pir Omar Gudrun. 24 Berossus names the mountains of the "Gordyaeans," or the Kurds, as the landing site. These mountains correspond with "Jebel Judi" in agreement with Syriac and Arabic traditions and lie in the southwestern part of Armenia. 25

Opening the Hatch

Genesis 8:6: "And it came to pass at the end of forty days, that Noah opened the window of the ark which he had made."

Utnapishtim: "*I opened the hatch, and the light fell upon my countenance. I was horrified, and I sat down and wept. Over my countenance ran my tears.*" 26

In the Gilgamesh legend, Utnapishtim waits for a "perfect" seven days after coming to rest, whereupon he is saddened at the sight. In the Genesis account, Noah waits a "perfect" forty days. No discrepancy here, both accounts use perfectly good numbers.

Sending Out the Birds

Genesis 8:7-12: "And he sent forth a raven, which went forth to and fro, until the waters were dried up from off the earth.

Also he sent forth a dove from him, to see if the waters were abated from off the face of the ground;

But the dove found no rest for the sole of her foot, and she returned unto him into the ark, for the waters were on the face of the whole earth: then he put forth his hand, and took her, and pulled her in unto him into the ark.

And he stayed yet other seven days; and again he sent forth the dove out of the ark;

And the dove came in to him in the evening; and, lo, in her mouth was an olive leaf pluckt off: so Noah knew that the waters were abated from off the earth.

And he stayed yet other seven days; and sent forth the dove; which returned not again unto him any more."

Utnapishtim: *"When the seventh day arrived, I sent forth and set free a dove. The dove went forth, but came back; since no resting place for it was visible, she turned round.*

Then I sent forth and set free a swallow. The swallow went forth, but came back; since no resting place for it was visible, she turned round.

Then I sent forth and set free a raven. The raven went forth and, seeing that the waters had diminished, he eats, circles, caws, and turns not round.

Then I let out (all) to the four winds ..." [27]

Xisuthros: *"Xisuthros let go some birds... But as they found no food nor a place to alight, they returned to the ship. After certain days Xisuthros again let the birds to; these again returned to the ship, but with their feet muddy. But when they were let go for the third time, they did not again return to the ship."* [28]

One thing is certain about human beings. Two people can read these accounts, and reach totally different conclusions. One person will observe that the striking similarities deny any possibility that these flood stories could have arisen from independent sources. Another may declare the Gilgamesh epic has the wrong birds in the wrong order, and makes no mention of an olive leaf. That is the nature of people, we are just that way.

In my pre-Christian, madcap days as a daring young aviator, I once attended a party in London and was afforded the opportunity to make a spectacle of myself. Two years later I was standing around telling "war stories" with my flying buddies, when one of them proceeded to tell a tale he had heard from some F-4 pilot.

Halfway through the story I realized who it was about. It actually sounded better the way it was being told, so that is one of the reasons I offered no corrections. And if that story is still making the rounds, it may sound even better today.

You see, I knew the story. Yes, it had been altered, rearranged, and embellished, but it was clearly recognizable. The same thing is true with these ancient legends compared to Genesis.

The narrative attributed to Utnapishtim cited above could be read in almost any church service, and unless members of the congregation were reading along in their Bibles, few would notice the difference. And those who would detect a dissimilarity might think it was being read from the Bible in another version.

The *Interpreter's Dictionary of the Bible* agrees on this point:

> There is, however, one flood tradition which bears such striking resemblance to the biblical story that it must be directly related to it. This is the cuneiform (Sumerian, Babylonian, Assyrian) tradition. [29]

The Sumerian, Accadian, and Assyrian accounts, as well as the inspired version in the Bible, are conspicuously related. Probably, they arose from one source initially, and went separate ways to end up in different books. The uninspired versions do not detract from Genesis, they corroborate Genesis. Young holds the same view:

Man would have handed that truth down to his descendants, and after the flood that truth would have been passed
on to those who were not in the line of promise as well as
to those who were in the line of promise. Among unbelievers we can well understand that the truth would become corrupted with superstition. [30]

Leonard Cottrell adds:

> The fact remains that there *was* a great flood. And it hap-
> pened in lower Mesopotamia, in the "Land of Shinar." [31]

Reflection on an Olive Leaf

Traditionalists who argue for a world-wide flood not only have
disregarded geological evidence, they have ignored the Bible's
evidence. Had the entire earth been submerged in salt water for
over nine months, plant life would have perished. Noah sent the
raven and dove to test the terrain, but the dove returned to the ark,
"for the waters were on the face of the whole earth ..." (Gen. 8:9).

Seven days later, "he sent forth the dove out of the ark," and
when it returned, "in her mouth was an olive leaf ..." (Gen. 8:10-11).
Could an olive tree survive over nine months underwater? If one
did, could it sprout leaves in a week? Or is it more sensible to
believe that most of the world, including much of Armenia from
where the leaf probably was taken, was spared the flood?

Altars, Offerings and Signs

Genesis 8:20: "And Noah builded an altar unto the Lord; and
took of every clean beast, and of every clean fowl, and offered
burnt offerings on the altar."

Utnapishtim: "*I poured out a libation on the top of the mountain. Seven and seven cult-vessels I set up, upon the pot-stands I
heaped cane cedarwood, and myrtle.*" [32]

<u>Ziusudra</u>: *"The king kills an ox (and) offers an abundant sacrifice of sheep."* [33]

An offering made at the end of the voyage is a conspicuous commonality in three of the five accounts.

<u>Genesis 8:21</u>: "And the Lord smelled a sweet savor;..."

<u>Utnapishtim</u>: *"The gods smelled the savor, the gods smelled the sweet savor."* [34]

In the end it is impossible to skirt the conspicuous similarities in both the inspired and uninspired versions, testifying to one memorable, but local, event.

The Token of a Covenant

<u>Genesis 9:13</u>: "I do set my bow in the cloud, and it shall be for a token of a covenant between me and earth."

The rainbow is a sign. It is unlikely God waited over 4 billion years before devising rainbows. Pilate did not invent the cross for the crucifixion of Christ. The Romans used the cross for executions before and after Christ died upon it. This does not diminish the cross as a sign for Christians, just as God used the rainbow as a sign for Noah.

Summary

Writing in 1683, over 150 years before the Sumerian and Accadian flood stories were unearthed, Matthew Poole had this to say in *A Commentary on the Holy Bible*:

> And whereas our modern heathens, that miscall themselves Christians, laugh at the history of this flood upon this and the like occasions, as if it were an idle romance; they may please to note, that their predecessors, the ancient and wiser heathens, have divers of them acknowledged the truth of it, though they also mixed it with their fables, which was neither strange nor unusual for them to do. [35]

Extra-biblical accounts help substantiate the flood as a documented event, an incident of record. The legends of Gilgamesh, Atrahasis, and Ziusudra not only establish the flood, they dictate the location, the extent, and the approximate date.

Adamite populations were the target of the flood. They resided in the heart of Southern Mesopotamia at that time, the focus point of the flood. The scope of the flood was entirely confined to this locale. Most of the world's human populations were unaffected. The time of the flood was around 2900 BC when Ziusudra was king.

If any credible evidence from nature of a world-wide flood catastrophe could be documented, could pass the scrutiny of peer review, and become published in any respected scientific journal, that discoverer would be an instant Noble Prize candidate. What has been offered up as evidence of a global cataclysm has been paltry, dubious, and unconvincing.

It has been pointed out that God's general revelation should match up with His special revelation. There are times when ignoring the clear messages from the Bible has caused some to rely too heavily on naturalistic revelation to the detriment of their conclusions. Likewise, God's messages from nature must be heeded to restrain us from making preposterous interpretations of Scripture. It serves no useful purpose to declare the Bible inerrant, and then interpret so erroneously that it causes disbelief.

NOTES

[1] M. E. L. Mallowan, "Noah's Flood Reconsidered," *Iraq*, n. s., 26 Part 2 (Autumn 1964), 70.

[2] Ibid., 72.

[3] Gleason Archer, *A Survey of Old Testament Introduction* (Chicago: Moody Press, 1974), 208.

[4] Lloyd R. Bailey, *Noah: The Person and the Story in History and Tradition* (Columbia: University of South Carolina Press, 1989), 32.

[5] James B. Pritchard, *Ancient Near Eastern Texts Relating to the Old Testament* (Princeton: Princeton University Press, 1955), 93.

6 Ibid., 44.

7 Ibid., 105.

8 Ibid., 93.

9 Ibid., 105.

10 Ibid., 93.

11 Ibid., 105.

12 Alexander Heidel, *The Gilgamesh Epic and Old Testament Parallels* (Chicago: The University of Chicago Press, 1963), 84.

13 Mallowan, "Noah's Flood Reconsidered," 72-75.

14 Albert T. Clay, *A Hebrew Deluge Story in Cuneiform* (New Haven: Yale University Press, 1922), 63.

15 Knut Tallquist, *Addadische Gotterepitheta* (Helsinki: 1938), 424-426.

16 Stephen Langdon, *Sumerian Liturgical Texts* (Philadelphia: University Museum, 1917), 85.

17 Pritchard, *Ancient Near Eastern Texts Relating to the Old Testament*, 94.

18 Ibid., 94.

19 Samuel Noah Kramer, *Sumerian Mythology* (New York: Harper & Brothers, 1961), 39.

20 W. F. Van Amringe, *An Outline of a New Natural History of Man* (New York: Baker & Scribner, 1848), 62.

21 Ibid., 62.

22 Clay, *A Hebrew Deluge Story in Cuneiform*, 79.

23 Heidel, *The Gilgamesh Epic and Old Testament Parallels*, 250.

24 Pritchard, *Ancient Near Eastern Texts Relating to the Old Testament*, 94.

25 Heidel, *The Gilgamesh Epic and Old Testament Parallels*, 250.

26 Clay, *A Hebrew Deluge Story in Cuneiform*, 78-79.

27 Pritchard, *Ancient Near Eastern Texts Relating to the Old Testament*, 94-95.

28 Heidel, *The Gilgamesh Epic and Old Testament Parallels*, 251.

29 *Interpreter's Dictionary of the Bible* (New York: Abingdon Press, 1962), Vol. II, 280.

30 E. J. Young, *In The Beginning* (Edinburgh: Banner of Truth Trust Publishers, 1976), 38.

31 Leonard Cottrell, *The Land of Shinar* (London: Souvenir Press, 1965), 133.

32 Pritchard, *Ancient Near Eastern Texts Relating to the Old Testament*, 95.

33 Heidel, *The Gilgamesh Epic and Old Testament Parallels*, 105.

34 Pritchard, *Ancient Near Eastern Texts Relating to the Old Testament*,
    95.
35 Matthew Poole, *A Commentary on the Holy Bible*, Vol. 1 (London: The
    Banner of Truth Trust, 1962) (orig. pub. 1683), 21.

Chapter 15

THE FLOOD WATERS ABATE: INTO THE PERSIAN GULF

In Genesis 1:25-27, the creation of man follows the creation of animal life. We discussed in chapter 11 that this order is reversed in Genesis 2:19-20. A possible reason for the reversal in order is that the animals created in Genesis 1 would cover the globe, whereas certain animals in Genesis 2 were indigenous to the garden of Eden. Adam was given the task of naming the animals. Not polar bears, penguins, aye ayes, South American tree sloths, Australian koala bears, or duck-billed platypuses, which would have been geographically remote, but simply the animals near to the garden. [1] It is the offspring of this group, apparently, that Noah was instructed to save in the flood.

All and Every

Genesis 8:17: "Bring forth with thee every thing that is with thee, of all flesh, both of fowl, and of cattle, and of every creeping thing that creepeth upon the earth; that they may breed abundantly in the earth, and be fruitful, and multiply upon the earth."

"Every beast" and "every fowl" that Adam named (Gen. 2:19) were the ones in this special area of habitation. The same requirement fits Noah's circumstances. "Every beast, every creeping thing, and every fowl" that Noah brought off the ark in Genesis 8:19 were likely from "all" the animals Adam named. They came from the immediate vicinity. We can only speculate on how much territory that would cover.

It is a great temptation to take ancient Hebrew words, translate them directly into English, and then make an interpretation based upon what modern English-speaking peoples might have meant had they used such words. There are many instances where this technique will generate an erroneous result.

In Genesis 41:41, 47, Pharaoh set Joseph "over all the land of Egypt," and there were seven plentiful years. "And he gathered up all the food of the seven years, which were in the land of Egypt ..." (Gen. 41:48). All the food? The resident Egyptians ate none of it in seven years?

"And the famine was over all the face of the earth ..." (Gen. 41:56). Were the Americas similarly affected? Australia? China? "And all countries came into Egypt to Joseph for to buy corn ..." (Gen. 41:57). That would be a long trip for someone living in Scandinavia.

Now, let us use common sense interpreting these verses. The issue is not whether Old Testament passages can be interpreted literally. They can and should be. There were seven years of bountiful harvest followed by seven lean years. Food was stored up during the first seven years so that enough would be available for the following seven. They were so efficient that even surrounding countries could draw on their stores.

Reason needs to be applied lest we cause a needless distortion. It would be unreasonable to suggest that the Egyptians ate not a morsel for seven years because "he gathered up all the food of the seven years." It would be senseless to think the rainfall in Peru was deficient because "the famine was all over the face of the earth," or that Aztec Indians lined up behind Australian Aborigines at the gates of Memphis because "all countries came into Egypt to Joseph for to buy corn...." By the same token, the Genesis flood narrative does not mandate a world-wide catastrophe because "all flesh died" in it.

This interpretative constraint can be seen elsewhere in the Old Testament. In I Samuel, David and 600 of his men were in hot pursuit of the Amalekite army. When David's band made contact with the Amalekites, "behold, they were spread abroad upon all the earth ..." (I Sam. 30:16).

Whereupon David smote them; only 400 young Amalekite men escaped death (I Sam. 30:17). To those who would insist that the language of Genesis 7 and 8 dictates a world-wide flood because the waters prevailed "upon the earth," I would invite them to be consistent, and distribute the Amalekite army over the globe also. Then explain how David was able to eradicate them in 24 hours with only 400 men (200 lagged behind).

The Amalekite army, in all likelihood, occupied no more territory than did the Confederate troops at the battle of Gettysburg. Knowing that gives us a means of measurement we can apply to the flood. By using the Bible's own yardstick, the deluge of Noah's day would be local, not global.

Another example of Hebrew terminology is found in Psalm 22. This is a psalm of David, yet a prophecy of the crucifixion, "... they pierced my hands and my feet" (Psa. 22:16).

Matthew harkens back to David, "the prophet," and quotes Psalm 22:18 in his account of the Roman soldiers casting lots for Jesus's garments (Matt. 27:35). Yet, David the psalmist also wrote, "... and all my bones are out of joint" (Psa. 22:14).

Should the word "all" in this verse cause heartburn? No! It is entirely consistent in Hebrew usage. Reason and common sense cannot be cast aside when Hebrew terminology is converted into English. Likewise, reason is helpful to recognize that the Genesis flood passages pertain assuredly to a local calamity, not a global cataclysm.

Men and Animals Spared

Placing every kind, variety, or species of animal, bird, and insect on a mountain top, or even on a lower foothill after the flood, would present a gigantic redistributional nightmare. Imagine tree sloths, for example, lumbering their way from Armenia up through Siberia, across the Bering Strait (long after the land bridge was gone), and then down through what is now Alaska and Canada, across the plains and deserts that are part of the U.S. and Latin America to their present-day home in South America. Whereupon, the giant-sized versions promptly went extinct! Drats.

If we know anything about world geography, we also should know that migration does not explain how animals could have traveled from Armenia to their present-day habitats. Before the flood, during the flood, and after the flood, the world's animal populations went about their daily business, oblivious to what transpired in Southern Mesopotamia.

The fact of animal survivors gives needed perspective to the biblical account. It may not be noticeable immediately from Genesis 8:17, but surviving animal life is one basis for acknowledging human survivors as well. It may not be as evident, but when we focus on what the Bible says, and not on what we have been told it means, we can see there were human populations living outside the flood zone.

Ramm emphasized this point:

> The flood was local to the Mesopotamian valley. The animals that came, prompted by divine instinct, were the animals of that region; they were preserved for the good of man after the flood. Man was destroyed within the boundaries of the flood; the record is mute about man in America or Africa or China. 2

If an abundance of animal life over the globe was excluded from the flood, something we can verify easily, then a consistent reading of the text also excludes mankind. Linking animals and man in the Genesis text requires a mutual interpretation. In ignorance, we might think all animals and all men perished in the flood. In light of general revelation, we can say that some animals and some men perished in the flood. It would be entirely inconsistent, however, to assert that only some animals died in the flood, but all men perished.

For Whom the Bell Tolled

In pointing to His second coming, Jesus refers to the days of Noah. "For as in the days that were before the flood they were eating and drinking, marrying and giving in marriage, until the day that Noah entered into the ark ..." (Matt. 24:38).

In Atrahasis, we are given a perspective on what "eating and drinking" may mean. Although there are pieces missing out of the account, enough has been recovered to show us the overwhelming compassion and sorrow he must have felt in the waning hours before the rain began to fall. After the birds, cattle, and wild animals were put aboard, Atrahasis turned to his people for whom there was no provision.

> He invited his people [   ]
> [   ] to a feast.
> [   ] he put his family on board.
> They were eating, they were drinking.
> But he went in and out,
> Could not stay still or rest on his haunches,
> His heart was breaking and he was vomiting bile. 3

As regards the Sumerians, "drinking" has a different connotation. Although some wheat was grown in Sumer, the salty, alkaline soil was more friendly to growing barley. The Sumerians knew what to do with that. Some 40% of all the barley grown in that region was used to produce ale. The drunken ways of the Sumerians were so notorious, the Greeks joked that one of their pagan gods, Dionysus, had fled from Sumer in revulsion. 4

The "wickedness" and "thoughts of evil" (Gen. 6:5) which brought on the ultimate destruction must have been manifested in those who had the knowledge and capacity for sin, specifically those who were direct descendants from Adam and Eve. Sumerians living in close proximity with the Semites could have been afflicted similarly by sin, and there is much evidence that the cancerous growth of sin had spread to them as well.

We know that slavery, divorce, and polygamy were practiced. The Sumerians worshipped over 3,000 pagan gods, and brought food offerings to them. As populations grew, appetites for more grain for food and drink put increased demands on the scarce water supply. When they dug irrigation canals upstream, it would deprive farmers farther downstream. Cities waged war on neighboring cities over land and water rights.

The Sumerian king list often concludes a list of kings at cities with an ominous phrase, "Uruk was smitten with weapons"; "Ur was smitten with weapons"; "Kish was smitten with weapons." [5] Although the Sumerians were capable of committing acts of raw aggression, murdering and enslaving their hapless victims, the question is, were they accountable?

God is a loving but righteous Father, meting out punishment for disobedience. Adam was given a commandment, he disobeyed, and was punished. When "men began to call upon the name of the Lord" (Gen 4:26), the knowledge of both good and evil passed to Adam's offspring.

Sin is a uniquely human attribute, but until God determined to give a commandment specifically to Adam, it does not appear that penalties were assigned. History does not record such a commandment was given to native Americans, or Black Africans, or Asiatics, or even Egyptians or Sumerians. They were not tasked by God. Why should He destroy those who were not held to account?

When Abraham made his appeal to God to withhold His judgment against the condemned city of Sodom if only ten righteous people were found, he started with this question in Genesis 18:23, "Wilt thou also destroy the righteous with the wicked?"

If God could confirm for Abraham that His judgment would not extend to punish the righteous few for the wicked many, then He merely read into the record what He had already established at the time of the flood. Was moral corruption prevalent among those who had yet to learn of sin and disobedience, who were not answerable, who had not been given a commandment? Probably so. Immorality - yes, judgment - no.

On the question, did Noah's flood cover the entire world? Donald Boardman answered "no," and concluded:

> There is little evidence from the Scriptures concerning how God was dealing with people in other parts of the earth. It seems logical in the light of these evidences that, in the case of the Noahic society, God was dealing with a local society and that his punishment was upon a limited number of persons at the time. [6]

Flood Survivors

If the flood is recent and local, and there are no gaps in the Genesis chronology which puts only 1,656 years (from the Masoretic text) between the flood and Adam's creation, then placing both Noah's flood and Adam's birth downstream in the history of humanity is a necessary conclusion. Not only unaffected animal survivors, but disinterested human survivors, were all over the globe when Noah disembarked. From *The Cambridge Ancient History*:

> Although the Flood was not the universal phenomenon that it has often been claimed to be, there is no doubt that it was exceptional among the long series of recorded Mesopotamian floods and that it overwhelmed parts of various cities in southern Babylonia. [7]

Dalley calls for survivors in her book, *Myths from Mesopotamia*:

> Although *Atrahasis* emphasizes the catastrophic nature of the Flood, the ancient Babylonians were well aware that not every thing was destroyed; *Erra and Ishum* makes it clear that the city of Sippar survived, a belief echoed by Berossus, who says that ancient writings were buried there before the Flood and later retrieved. [8]

Writing at the time of Caesar Augustus, Nicolaus Damascenus makes mention of one who was carried on an ark to the mountains of Armenia. His brief account ends with a conjectural comment, "Perhaps this was the same individual of whom Moses the legislator of the Jews has made mention." [9] (Hmmm, I wonder ...)

It is the beginning of Damascenus's account that bears directly on the question of local flood survivors.

> There is above Minyas in the land of Armenia a very great mountain which is called Baris; to which, it is said, that many persons retreated at the time of the deluge and were saved ... [10]

When we consider that racially diverse populations covered the globe long before 5,000 years ago, flood survivors are mandated. [11] In his book, *The Biblical Flood*, Davis Young concluded:

> ... archaeological investigations have established the presence of human beings in the Americas, Australia, and southeastern Asia long before the advent of the sort of Near Eastern civilization described in the Bible and thus long before the biblical deluge could have taken place. In the light of a wealth of mutually supportive evidence from a variety of disciplines and sources, it is simply no longer tenable to insist that a deluge drowned every human on the face of the globe except Noah's family. [12]

"All the relevant evidence from the created order tells us that the flood was neither geographically nor anthropologically universal," [13] Young went on to say. Indeed, the Bible itself appears to be cognizant of flood survivors. The Genesis 6:4 "giants" (*Nephilim* in the Hebrew) were some manner of men with ancient origins who apparently were in residence prior to Noah, and maybe, Adam. Furthermore, they appear in later chapters. In Numbers 13:33, the post-flood "sons of Anak who come of the giants," reflects back to Genesis 6:4, to the pre-flood period. This is from *The Expositor's Bible Commentary*:

> On the face of it, the remark presents a problem to the view that only Noah and his sons survived the Flood, since it suggests that the "sons of Anak" were descendants of the "Nephilim" (*min hannepilim*, lit. "from the Nephilim") who lived before the Flood. [14]

How could the Nephilim be on both sides of the flood? Because in the post-flood period they were living in what became Canaanite country, the region of Palestine, outside the flood zone. Data compiled from archaeological excavations in the Near East corroborates the local nature of the flood. Although flood deposits have been found at sites in Mesopotamia, no flood deposits have been found in Egypt or in the Palestine region.

In Deuteronomy 2:10, 11, "The Emims dwelt therein in times past, a people great and many, and tall as the Anakims; which also were accounted giants, as the Anakims; but the Moabites call them Emims." The Anakims were a race of giants, descendants of Anak, who dwelled in southern Canaan. [15] Emims were as tall as the Anakims, the Bible attests, and were the ancient inhabitants of Moab. [16] In Deuteronomy 2:20, 21, "That also was accounted a land of giants: giants dwelt therein in old time; and the Ammonites call them Zamzummims; a people great and many and tall as the Anakims; but the Lord destroyed them before them; and they succeeded them, and dwelt in their stead." Joshua mentions "remnant of the giants," "giants," or "valley of the giants" in five verses (Josh. 12:4; 13:12; 15:8; 17:15; 18:16).

Post-flood Emims, Anakims, or Zamzummims cannot be identified as Ubaidans, Sumerians, or Persians, but likewise, they do not appear to be any of Noah's kin either. And this urges the question. How could there be populations unrelated to Noah, giants or otherwise, if all men died in the flood? If the Bible has no problem with flood survivors, it should not bother us.

Only one year after Darwin published *The Origin of Species*, fellow Englishman Edward William Lane wrote *The Genesis of the Earth and of Man.* In it he said:

> It appears, therefore, that Holy Scripture does not forbid, nay, rather it requires, a belief in the existence of Pre-Adamites of our species, whose posterity were not destroyed with the unbelieving Adamites by the waters of the flood. [17]

If Bible scholars had paid as much attention to Lane as biologists did to Darwin it is possible that no Bible-science conflict would have developed at all!

Of Patriarchs and Kings

When the Sumerian king lists began to surface, there was a rush to show that these were the source of the biblical patriarchs.

The Berossus list, close companion to the Sumerian versions, was analyzed by the Assyriologist Zimmern, who concluded:

> It can hardly be doubted that the Biblical tradition of Gen 5 (P) concerning the antediluvian patriarchs is basically identical with the Babylonian tradition about ten antediluvian primeval kings. [18]

Taking the opposite tack, G. F. Hasel made a comparative study and found, "a complete lack of agreement and relationship" [19] between Genesis 5 and 11 and the Sumerian kings. As is often the case, the truth may be found between extremes. The Genesis patriarchs and Sumerian kings cannot possibly be "basically identical" for reasons we shall see. On the other hand, there is sufficient commonality that to say there is "a complete lack of agreement" is equally erroneous. We will compare by first deriving a concensus list of Sumerian kings.

Table 1

A Revised King List

W-B 444	W-B 62	Revised King List	City
1 Alulim	1 Alulim	1 Alulim	Eridu
2 Alalgar	2 Alalgar	2 Alalgar	"
3 Enmenluanna	3 -kidunnu	3 Enmenluanna	Badtabira
4 Enmengalanna	4 -alimma	4 Enmengalanna	"
5 Dumuzi	5 Dumuzi	5 Dumuzi	"
6 Ensibzianna	6 Ensipazianna	6 Ensipazianna	Larak
	7 Enmenluanna		
7 Enmenduranna	8 Enmenduranna	7 Enmenduranna	Sippar
8 Ubardudu	9 Ubartutu	8 Ubartutu	Shuruppak
		9 Suruppak	"
	10 Ziusudra	10 Ziusudra	"

--- **The Flood** ---

Using a Revised King List

So that we may use only one list for comparison, King list W-B 62 is revised, taking into account another primary list (W-B 444), plus five other lists (not shown). Table 1 is the result.

Step 1. Misplacing names was a common scribal error. With the other lists as guides, Enmenluanna replaces the fragmented -kidunnu, moving him from seventh to third. This moves the eighth and ninth kings up to positions 7 and 8.

Step 2. Enmengalanna from W-B 444 replaces -alimma.

Step 3. Suruppak is inserted at position 9 as an intermediate generation. Ziusudra's grandfather, Ubartutu, reigned immediately before Ziusudra, but Suruppak was Ziusudra's father. [20] One king list even names Suruppak and omits Ubartutu.

We now have a revised king list for comparison purposes.

Table 2

Pre-Flood Patriarchs And Kings

Patriarch	Rev. King List	Berossus	W-B 444	UCBC 9-1819
1 Adam	1 Alulim	1 Alorus	1 Alulim	1 Alulim
2 Seth	2 Alalgar	2 Alaparos	2 Alalgar	2 Alalgar
3 Enosh	3 Enmenluanna	3 Amelon	3 Enmenluanna	3 Ammeluanna
4 Cainan	4 Enmengalanna	4 Ammenon	4 Enmengalanna	4 Ensipazianna
5 Mahalalel	-----	5 Megalaros	-----	-----
-----	5 Dumuzi	6 Daonus	5 Dumuzi	5 Dumuzi
6 Jared	6 Ensipazianna	8 Amempsinos	6 Ensibzianna	-----
7 Enoch	7 Enmenduranna	7 Edoranchus	7 Enmenduranna	6 Enmenduranki
8 Methuselah	8 Ubartutu	9 Otiartes	8 Ubardudu	7 Ubartutu
9 Lamech	9 Suruppak			
10 Noah	10 Ziusudra	10 Xisuthros		8 [Ziusudra?]

--- **The Flood** ---

Observations

Table 2 gives us a "spreadsheet" of the pre-flood patriarchs, including the revised list of pre-flood kings, the Berossus list, and two other king lists to compare alongside the Genesis patriarchs.

One transposition has been performed on the Berossus list. Both Amempsinos and Ensibzianna are identified as king of Larak. Since Larak was "clearly the third city" according to Langdon, [21] this suggests the Berossus list has Amempsinos out of order with Edoranchus.

Let us start with some preliminary observations. First, the genealogies in Genesis are just that: the early fathers of the Adamites and the Semites. The Sumerian king lists represent Semite (Adamite) and Sumerian kings, although there is some disagreement among experts as to which is which. At any rate, as the king lists intend to represent rulers, no purely ancestral relationships are implied, even though royal offspring often ascend the throne.

Second, the thousands of years the pre-flood kings reigned looks to be an error in interpretation rather than a recording error. This can be deduced from the post-flood kings at Kish. After "the flood swept thereover," and the kingship was restored, 23 kings reigned a total of 24,510 years - plus, if you can believe it, 3 months and 3 1/2 days! (Archbishop Ussher must have had a Sumerologist counterpart.)

Using the archaeological date of 2900 BC for the flood, that would mean the kings of Kish are still ruling today, and have another 19,000 years to go! Where is the error? The years the post-flood Sumerian kings reigned appear to be off by a factor of about 60. The Sumerians used a sexagesimal system of numbers, and that offers a clue as to how astronomical figures may be brought into the realm of believability. Dividing by 60 puts the total years reigned at Kish at a little over 400, a reasonable figure. It can get more complex than that (they may have relied on moon phases rather than sun cycles, etc.), but it's not something we need to dwell on.

To assert that the Bible genealogies are unrelated to the Sumerian kings because of a discrepancy in the hundreds of years of life for the patriarchs, versus the thousands of years reigned for the pre-flood kings, misrepresents the case. It should not be surprising that Sumerologists have been every bit as prone to error as Bible translators, and similarly reluctant to make corrections. [22]

Third, confusion can arise when more than one name pertains to a single individual. Among the difficulties is that titles or occupations have been used at times, rather than proper names, and will look dissimilar, especially when recorded in different languages. There are many instances where the Bible itself uses more than one name for one person, for example: Abram = Abraham, Jacob = Israel, Saul = Paul, Peter = Simon = Cephas, and even Jesus = Emmanuel (corresponding, perhaps, to the Accadian "Ea").

Fourth, Adam is a prime candidate as Alulim at Eridu. Seth, or conceivably Enosh, could be the second king, Alalgar. But the fourth patriarch, Cainan, does not and should not appear on the king lists. Eridu was overthrown. Kingship passed to the victorious city - a Sumerian city - Badtabira. A Sumerian city at that early date was likely devoid of foreigners speaking strange languages. The three kings of Badtabira should not be in the Adamic line.

So a dissimilarity is what we should expect concerning those three kings, and that is the case. Also, no connection can be seen between any of the kings and Jared, or with Mahalalel outside of Berossus. This sets apart at least three or four out of the ten patriarchs as absent from the Sumerian king lists, and that is about as far as dissimilarity can be extended.

Finally, there are complicating factors. The genealogies are in Hebrew, while the kings are in Sumerian, an unrelated language, and Berossus wrote in Greek. Still, these are not insurmountable obstacles. In Table 3 we will see the list of patriarchs and the lists of kings are not completely independent - there is a relationship; not a one for one relationship, but an obvious relationshp nonetheless.

Table 3

Comparison Of Pre-Flood Patriarchs With Revised King List

Patriarch	Rev. King List	City	Relationship
1 Adam	Alulim	Eridu	Probably is the same man
2 Seth	Alalgar	"	Conceivably the same man
3 Enosh	Enmenluanna	Badtabira	Most probably different men
4 Cainan	Enmengalanna	"	Should be different men
5 Mahalalel	Dumuzi	"	Virtually have to be different
6 Jared	Ensipazianna	Larak	No similarity seen
7 Enoch	Enmenduranna	Sippar	Quite possibly the same man
8 Methuselah	Ubartutu	Shuruppak	Should be the same man
9 Lamech	Suruppak	"	Likely the same man
10 Noah	Ziusudra	"	Essentially is the same man

--- The Flood ---

Line-by-line Explanation

Line 1. A similarity between Alulim and Adam can be seen just as the Accadian father-god *Ilu* foreshadows the Canaanite *El*. Parallels between the Sumerian Alulim, the Accadian Adapa, and the Hebrew Adam point toward a commonality. Clay proposed that Alorus from the Berossus list was "El-Or" found in early Aramaic inscriptions [23] - and therefore, a Semitic (Adamic) name. Who would have been the first father or king of the forerunners to the Semites if not Adam? And if Adam, special in many respects, resided in Eridu from the start, who better to serve as king?

Line 2. Some scholars make the connection: Alaparos = Adapa = Adam, making Adam the *second* king. This raises a question. If Adam was the second king, who was the first? It seems equally reasonable to suggest that Seth, or one of Adam's other sons, or even Enosh could have been this monarch.

Alalgar may have been one of Adam's offspring. There is no way of knowing, but Poebel credits Berossus's Alaparos as the "son of" Alorus. [24] Furthermore, if the first king at Eridu was Adam, a non-Sumerian, the next king, if directly related, would also have been non-Sumerian. Keep in mind, the first two names, Alulim and Alalgar, are Semitic (or Adamic), not Sumerian names.

The Semitic (Adamic) name, Alalgar, is entirely appropriate as applied to the covenant family. Among the meanings offered for Alaparos are "Ox of the god Uru," and "Lamb of El." [25] "El," Assyrian for God (and seen in Hebrew as *Elohim, El Shadai*), was the father god, first in the early Accadian trinity. Thus, the name could be rendered literally, "Lamb of God." This description of profound theological significance used of Jesus (John 1:29, 36) might have been applied to Seth, or even Enosh, when men began "to call upon the name of the Lord" (Gen. 4:26). Seth, one of his brothers, or his son may have been this second pre-flood king.

Line 3. Alalgar's rule was closed out when Eridu was overthrown and kingship passed to the victorious Enmenluanna, king of Badtabira, a Sumerian city. It would be unlikely that one of Adam's immediate generations (for example, Enosh) would have made war on his own father or grandfather. Also, Enmenluanna is a Sumerian name, making him the first genuine Sumerian on the Sumerian King List. It follows that a non-Adamic ancestry would be implied for this Badtabiran king and his successors.

Considering Adam's longevity, 930 years, he and at least some of his kin must have escaped the bloodshed at Eridu. If Adam moved north about 50 miles to Erech, adjacent to Enoch, the city Cain built, this would have brought Adam to a location where he and his entourage could find refuge and safety among family members.

Line 4. From the name Enmengalanna, we might suspect he was son and successor to the throne of Enmenluanna. Adamic ancestry is equally unlikely therefore, and is reflected by a dissimilarity between him and the fourth patriarch, Cainan.

Line 5. In his analysis, Clay allowed, "It seems that Mahalal-El may be represented by Megalaros ..." [26] A link between Mahalalel and the fifth king on the Berossus list looks credible. But probably, he is not the fabled Dumuzi who corresponds to Daonus, sixth on the Berossus list, as Dumuzi and Daonus are both identified as "a shepherd" and "the shepherd." [27] Dumuzi was consort to Inanna, "queen of heaven and earth." W-B 444 offers no additional data on any of its kings with one exception, declaring Dumuzi "divine," and his vocation as "the shepherd." [28] "Tammuz," the Semitic name for Dumuzi, [29] was famous in Accadian literature with a cult following to rival that of Elvis today.

In the Accadian legend, Adapa gained entrance to heaven by flattering Tammuz. "At the gate of Anu," Adapa told Tammuz how much he was missed on earth. [30] A thirty-eight line liturgical hymn to the departed Tammuz "represents the people wailing for the lord of life who now sleeps in the lower world." [31]

And the prophet Ezekiel had a vision where he was "brought to the door of the gate of the Lord's house," and "there sat women weeping for Tammuz" (Eze. 8:14). Thus, because of the cult following, the prophet Ezekiel bestowed biblical recognition on the celebrated Dumuzi, the fifth Sumerian king.

Line 6. Demonstrating that kingships were temporary and easily terminated in the land of Sumer, "kingship passed to Larak" when Badtabira was overthrown, and Ensipazianna became king. [32] It is doubtful that Jared, sixth in the line of patriarchs, could have been king of Larak, almost assuredly an entirely Sumerian city at that early date.

Line 7. "Sevens" often indicate that something may be unusual or important. Here may be another example. In Clay's words, "This king (Enmenduranna) is generally regarded as the original of the biblical Enoch." [33] We might argue what he meant by "original," but a commonality can be seen. Berossus has "Edoranchus," so all lists show a similarity.

Enmenduranna is deemed identical with Enmeduranki, sage and king of Sippar. [34] Zimmern, who first made the identification, said the name was pronounced "Evvedoranki." "Evved or Eved suggests the Hebrew *'Ebhed*, Clay contends. [35] If so, this could indicate Adamic ancestry for the king of Sippar who, according to Sumerian legend, was taken by the gods and taught divine mysteries. [36] And, "By faith Enoch was translated [taken up] that he should not see death" (Heb. 11:5).

Another consideration is that Sippar was the cult center of the sun god. The sun completes a cycle every 365 days, corresponding to Enoch's 365 years. [37] If Enoch was the king of Sippar who wrested power from Larak control, and then was taken by God, a void would have been left in the kingship. Or perhaps, someone not of good standing took his place. Either way, "Sippar was overthrown, its kingdom passed to Shuruppak." [38]

Line 8. The next three men on the revised list lived at Shuruppak before the flood. The Sumerian records show a direct line of descent from the king of Shuruppak, Ubartutu, through his son Suruppak to the last pre-flood king, Ziusudra. Ubartutu was Ziusudra's grandfather, while Noah's grandfather was Methuselah. Are Methuselah and Ubartutu one and the same?

W-B 62 ends in Ziusudra, although from W-B 444, only "one king reigned" at Shuruppak. [39] This was Ubartutu. If Ubartutu is Methuselah, who died near the time of the flood, this could explain the discrepancies in the two king lists. One list (W-B 62) recognizes Ziusudra who, if he ruled at all, reigned for less than a year, or at most only a few years before the flood. The other list (W-B 444) gives him no credit for an abbreviated rule at Shuruppak.

Line 9. Lamech begat a son: "And he called his name Noah ..." (Gen. 5:29). "With a brilliant name, let me make you famous," Suruppak told his son Ziusudra. [40] If Noah and Ziusudra are the same person, then unless he had two fathers, Lamech, the ninth patriarch, should be synonymous with Suruppak. A reason Suruppak never reigned could have been because his father outlived him. And Methuselah outlived Lamech.

Line 10. There is no need to recite the accomplishments of Noah. The names may not look alike, gift-wrapped in different languages, and touching on different facets of the man: "he who laid hold on life of distant days" (Ziusudra); "he saw or found life" (Utnapishtim); "the exceeding wise" (Atrahasis); and "rest or comforter" (Noah). [41] But corresponding flood stories using these names, recorded in Sumerian, Accadian, and Assyrian, all parallel the biblical deluge. These remarkably similar accounts would be hard to attribute to more than one man. Ziusudra, Atrahasis, Xisuthros, Utnapishtim, and Noah all seem to equate.

What Does It All Mean?

After a detailed analysis of Berossus, Delitzsch agreed with Zimmern and concluded:

> The ten Babylonian kings who reigned before the Flood have been accepted in the Bible as the ten antediluvian patriarchs, and the agreement is perfect in all details. [42]

What Delitzsch failed to recognize is that agreement could be expected only in instances where patriarchs were rulers, or conversely, when the kings were also in the covenant line from Adam. Evidently, some of the patriarchs did reign over small kingdoms. Yet, concurrent kingdoms were also established in Southern Mesopotamia ruled by non-biblical monarchs. Clearly, it was the intent of Berossus and the king lists to record a sequence of kings without regard to ancestry, just as it was the Bible's intention to record a certain line of ancestry whether or not they were kings.

In Sumerian, the first two letters *En-* of a ruler's name denotes kingship similar to the way we use "lord" in English. The god "Enki" combines *en* for "lord" and *ki* for "earth" to mean literally, "Lord of the Earth." The Sumerian word *lil* can mean "air," "breath," or "spirit." [43] Enlil was second in the Sumerian pantheon after the father god, An. The possible interpretations of this name should be obvious. A parallel could exist between this Sumerian and Accadian god and our Holy Spirit.

If we survey the pre-flood fathers, in both the line of Seth and the line of Cain, we see "En-" as the first two letters more often than any other combination (Enosh, once and Enoch, twice). It is quite possible, then, that both Cain's son and Seth's son were rulers over Sumerian subjects. This offers another clue that the seventh patriarch, Enoch, was also a ruler.

One final thought. The Bible submits no data whatsoever on seven of the ten pre-flood patriarchs beyond their age when the first son was born, age at death, and that they had "other sons and daughters." Details beyond that are given for only three: Adam, Enoch, and Noah. And the supplementary biblical information provided for each of them correlates directly to Sumerian and Accadian narratives.

Likewise, in all the Sumerian king lists pertaining to the pre-flood era, additional particulars are given on only one man, "divine Dumuzi, a shepherd." And he is the only Sumerian king, outside the line of Adam, corroborated in the Bible by his Semitic equivalent, "Tammuz." All coincidence, do you suppose?

Summary

Noah's Flood, recent in occurrence and confined to the Mesopotamian valley and its inhabitants, was retribution for sin, but as Paul states, "Sin is not imputed when there is no law" (Rom. 5:13b). Those civilizations outside the Adamic covenant and outside the immediate area were unaccountable and unaffected by the flood. If we take into consideration the allowable interpretations of "earth" instead of "land," "heaven" rather than "sky," and "mountains" as against "hills," coupled with the Hebrew words "all" and "every" when we would say "much" and "many," plus the Hebrew penchant for perfect or prophetic numbers, we should be able to understand how a Mesopotamian calamity has been misunderstood as a global cataclysm.

The biblical, archaeological, and anthropological evidence corroborates that God spared human populations who were outside the Mesopotamian valley and outside of His covenant. God "winked at" their ignorance (Acts 17:30), but targeted the Adamites in particular, obliterating those who were answerable and willfully disobedient. Evidently the Sumerians were hapless bystanders, many of whom perished, and some may have become proselytes who drowned in the flood.

In Luke, the Pharisees asked Jesus to rebuke His disciples, "And He answered and said unto them, I tell you that, if these should hold their peace, the stones would immediately cry out" (Luke 19:40). The "stones" in the form of inscribed clay tablets are crying out today, confirming God's Word. Are we listening, or are we like the Pharisees?

NOTES

1 Gleason L. Archer, *Encyclopedia of Bible Difficulties* (Grand Rapids: The Zondervan Corporation, 1982), 60.

2 Bernard Ramm, *The Christian View of Science and Scripture* (Grand Rapids: William B. Eerdmans Publishing Co., 1954), 169.

3 Stephanie Dalley, *Myths from Mesopotamia* (New York: Oxford University Press, 1989), 31.

4 George Constable, ed., *The Age of God-Kings: TimeFrame 3000-1500 BC* (Alexandria: Time-Life Books, 1987), 27.

5 Thorkild Jacobsen, *The Sumerian King List* (Chicago: University of Chicago Press, 1939), 93-105.

6 Donald Boardman, "Did Noah's Flood Cover the Entire World?" *The Genesis Debate*, ed. Ronald Youngblood (Grand Rapids: Baker Book House, 1990), 227.

7 I. E. S. Edwards, C. J. Gadd and N. G. L. Hammond, eds., *The Cambridge Ancient History* Vol. I, Part 2 (Cambridge: Cambridge University Press, 1971), 243.

8 Dalley, *Myths from Mesopotamia*, 6.

9 Isaac Preston Cory, *Ancient Fragments of the Phoenician, Chaldean, Egyptian, Tyrian, Carthaginian, Indian, Persian, and Other Writers* (London: William Pickering, 1832), 49.

10 Ibid., 49.

11 Roger Lewin, *In the Age of Mankind* (Washington: Smithsonian Books, 1988), 206-225.

12 Davis A. Young, *The Biblical Flood* (Grand Rapids: William B. Eerdmans Publishing Company, 1995), 242.

13 Ibid., 242.

14 Walter C. Kaiser Jr., Bruce K. Waltke and Ralph H. Alexander, eds., *The Expositor's Bible Commentary* (Grand Rapids: Zondervan Publishing House, 1990), 79.

15 From the Hebrew word, *'Anaqiy* meaning "long-necked."

16 From the Hebrew word, *'Eymiym* meaning "terrors." 1991.

17 Edward William Lane, Reginald Stuart Poole, ed., *The Genesis of the Earth and of Man* (London: Williams and Norgate, 1860), 103-104.

18 H. Zimmern, *Urknige und Uroffenbarung* (Gttingen: Vandenhoeck und Ruprecht, 1902), 539.

19 G. F. Hasel, "The Genealogies of Genesis 5 and 11 and their Alleged Babylonian Background." *Andrews University Seminary Studies*, n.s., 16 (Autumn 1978), 361-74.

20 Bendt Alster, *The Instructions of Suruppak* (Copenhagen: Akademisk Forlag, 1974), 43-49.

21 Stephen Langdon, *Oxford Edition of Cuneiform Texts*, Vol. II (London: Oxford University Press, 1923), 2-3.

22 One method of reconciling ages of Sumerian kings is outlined in an article by Hildegard Wiencke-Lotz, "On the Length of Reigns of the Sumerian Kings," *Chronology and Catastrophism Review, Journal of the Society for Interdisciplinary Studies* (vol. XIV August 1992), 20.

23 Albert T. Clay, *The Origin of Biblical Traditions* (New Haven: Yale University Press, 1923), 131.

24 Arno Poebel, *Historical Texts* (Philadelphia: The University Museum, 1914), 85.

25 Clay, *The Origin of Biblical Traditions*, 132.

26 Ibid., 135.

27 Langdon, *Oxford Editions of Cuneiform Texts*, Vol. II, 3.

28 George A. Barton, *The Royal Inscriptions of Sumer and Akkad* (New Haven: Yale University Press, 1929), 347.

29 Samuel Noah Kramer, *Myths of Enki The Crafty God* (New York: Oxford University Press, 1989), 7.

30 Stephen Langdon, *Sumerian Liturgical Texts* (Philadelphia: University of Pennsylvania Museum, 1917), 42.

31 Ibid., 285.

32 Barton, *The Royal Inscriptions of Sumer and Akkad*, 347.

33 Clay, *The Origin of Biblical Traditions*, 135.

34 Ibid., 135.

35 Ibid., 136.

36 Alexander Heidel, *The Gilgamesh Epic and Old Testament Parallels* (Chicago: The University of Chicago Press, 1963), 141.

37 Lloyd R. Bailey, *Noah: The Person and the Story in History and Tradition* (Columbia: University of South Carolina Press, 1989), 125.

38 Barton, *The Royal Inscriptions of Sumer and Akkad*, 347.

39 Weincke-Lotz, "On the Length of Reigns of the Sumerian Kings," 22.

40 Alster, *The Instructions of Suruppak*, 43.

41 Heidel, *The Gilgamesh Epic and Old Testament Parallels*, 227.

42 Frederich Delitzsch, *Babel and Bible* (Chicago: The Open Court Publishing Company, 1906), 41.

43 Samuel Noah Kramer, *History Begins at Sumer* (Philadelphia: The University of Pennsylvania Press, 1981), 76.

Chapter 16

THE POST-FLOOD PERIOD: OUT AND ABOUT

One of the major themes of this book is that human rationality has wrought a terrible toll on the biblical text causing needless tension between science and the Bible. Another example of this concerns the post-flood period migrations of Noah's three sons. Traditionalists have attributed all the present-day inhabitants of the globe solely to the wanderlust of Shem, Ham, and Japheth and their descendants.

The languages of the Noahic tribes were scrambled at the tower of Babel, as the explanation goes, and off they went to become Eskimos, Aztecs, Norsemen, Pygmies, Apaches, Laplanders, Chinese, Japanese, and so forth. No matter how much hot-air rhetoric is exhausted on this notion, world history will not capitulate, and neither will the Bible.

Unwelcome Rearranging

When my daughter was three years old, already she understood a biblical truth that has escaped the attention of many Bible commentators. That might sound like a bragging parent, but consider: "Thou hast hid these things from the wise and prudent, and hast revealed them unto babes" (Matt. 11:25).

What could a child of three possibly know that grown-up authorities on the Bible would not know? She knew Genesis 10 comes before Genesis 11. Now, I admit I kept it simple by just asking her to count to 20, and truth to tell, she sometimes got confused after 15, but rarely did she make a mistake between 1 and 12. My guess

The Origins Solution

is that my daughter is not too unusual. Most grade school children, and even high school students too, could list the chapters of any Book of the Bible in correct chronological order if they just counted in the normal manner.

A condition we might call biblical dyslexia does not appear to set in until seminary school. It is here and in Christian colleges all over America that human rationality suppresses what we learned and trusted as youngsters. The overpowering tendency has been to put Genesis 11 ahead of Genesis 10 so that everyone can get their tongues confused at Babel before they dispersed and took off for distant lands. This is a mistake that has heaped confusion on top of confusion needlessly.

The order of presentation in the first eleven chapters of Genesis requires no rearranging for enhanced palpability. For example, critics of Bible integrity have pointed to the fact that the order of creation in Genesis 1 is different from Genesis 2, changing from animals first, and then man - to man, and then animals. We saw a reason for that. Adam had to name only those animals God created expressly for his pleasure in Genesis 2. In turn, Noah had to save the descendants from this smaller group, and was thus spared the toil of building a bigger ark.

Here again, the Bible has it right and the commentators have it wrong. The dispersion of Noah's sons with their children and flocks in Genesis 10 comes before the confusion at Babel in Genesis 11 primarily because the post-flood dispersion happened at an earlier point in history than the tower of Babel incident. It doesn't get any simpler than that.

Dispersing the Nations

Genesis 10 announces the dispersion of Shem, Ham, and Japheth. Nimrod led at least one Hamitic group that helped rebuild cities, including Babylon. Asshur, a Semite, led an expedition and set out to build cities up north, assisted or resisted by the local populations. Other Semites set out as recorded in Genesis 10:30. The dispersion of the nations emanating from the three sons of Noah is completed in Genesis 10. The next event of importance takes place in the land of Shinar described in the next chapter of Genesis, discussed in the following chapter in this book.

The Bible negates the idea that all of us could have descended from Noah by the way the dispersion is narrated. The scattering of the sons after the flood is not recorded as a global intrusion. From the description alone we can see how far Noah's offspring ventured.

Genesis 10:1: "Now these are the generations of the sons of Noah, Shem, Ham, and Japheth: and unto them were sons born after the flood."

The verses following Genesis 10:1 list Noah's sons and grandsons. The pattern blurs a bit as some descendant lines are named to the third or fourth generation, and some are named beyond that. The names of some tribes are included; presumably they derived from the founding patriarchs of those generations. Later, lands that were settled by these tribes bore the same names, such that new immigrants would take on the name of the land, whether they were descendants or not.

Through further biblical references, and aided by history and archaeology, we have some idea of the migrations of these peoples and tribes. Please keep in mind, Bible scholars do not all agree on the eventual homelands of Noah's progeny. For our purposes, the Keil and Delitzsch Commentary is the primary reference, a scholarly work, and typical of what could be found in other Genesis commentaries.

Sons of Japheth

Genesis 10:2. Japheth had seven sons: "Gomer, Magog, Madai, Javan, Tubal, Meshech, and Tiras." Gomer may have started the tribe of the *Cimmerians* from whom came the *Cumri* or *Cymry*, some of whom migrated later to Wales and Britanny. Josephus attaches Magog to the *Scythians* on the Sea of Asof and the Caucasus.

The *Medes* are derived from Madai. Javan is given credit for founding the Greeks. Herodotus placed descendants of Tubal on the east of Thermodon. Meshech relates to the Moschi "in the Moschian mountains between Iberia, Armenia, and Colchis." [1] Tribal offshoots of Tiras have been associated with Thrace, Egypt, and Assyria.

Genesis 10:3. Gomer had three sons: "Ashkenaz, Riphath, and Togarmah." Ashkenaz has been linked with some peoples in the area of Germany, and possibly with the *Ascanians* in upper Phrygia. The Celts may be offshoots of Riphath, whereas Josephus reckoned Paphlagonia as the land of choice, but neither of these may be accurate. The Armenians can be traced to Togarmah.

Genesis 10:4. The sons of Javan were "Elishah, Tarshish, Kittim, and Dodanim." Elishah and his tribe may have headed for Sicily or Thessalonica. Spain may have been the land of choice for the offshoots of Tarshish, although Tarsus in ancient Cilicia, birthplace of Paul, is *Tarshish* in Hebrew. Kittimites can be found in Cyprus, and possibly, on the shores at the eastern end of the Mediterranean Sea. Dodanim's descendants are connected with the north of Greece.

Genesis 10:5: "By these were the isles of the Gentiles divided in their lands; every one after his tongue, after their families, in their nations."

The generally accepted interpretation of this verse is that the Japhethites, named in the previous verses, settled in the immediate vicinity of the Mediterranean Sea, with the possible exception of some who may have worked their way into parts of northwestern Europe.

The "islands" mentioned here and elsewhere in the Old Testament are the "islands and coastlands of the Mediterranean, on the European shore from Asia Minor to Spain." [2] The prophet Ezekiel accounts for the tribes of four of the seven sons of Japheth, and implies local confines.

The Lord spoke to the prophet in Ezekiel 38:2-6. In God's words, "Behold, I am against thee, O Gog, the chief prince of Meshech and Tubal: And I will turn thee back, and put hooks into thy jaws, and I will bring thee forth, and all thine army, horses and horsemen, all of them clothed with all sorts of armor, even a great company with bucklers and shields, all of them handling swords: Persia, Ethiopia, and Libya with them; all of them with shield and helmet: Gomer, and all his bands; the house of Togarmah of the north quarters, and all his bands: and many people with thee" (Eze. 38:3-6).

Take particular note of the last sentence, "and many people with thee." If the progeny of the sons of Noah accounted for all the peoples of the earth, then who were these "many people" who apparently came from outside the tribes of Japheth? Also, the lands named (Persia, Ethiopia, and Libya) are not far removed from Israel and Judah, or Babylon where the prophet was a captive resident at the time of his writing. None was called forth from the lands of Borneo or Brazil, for example.

It should be evident that the areas mentioned by historians, as well as the prophet Ezekiel, stop short of defining the entire globe. The traditional approach promulgated by conservative Bible apologists has been that the Japhethites were the seeds sown which led to the population of Europe and parts of Asia. That could only have been possible had they started their migrations many centuries before 2900 BC.

By this time in history, the continent of Europe and the British Isles were sparsely settled, including the southern part of Scandinavia. The Japhethites arrived on the scene too late in human history to account for the origins of the many diverse populations already residing in those parts who were busy cranking out pottery, stone carvings, and making megalithic tombs of giant slabs of stone that still can be seen today.

Just because pockets of populations can be found scattered about that can be traced back to the named patriarchs does not by any stretch of the imagination imply that all existing populations can be so traced. The migrating bands of Japhethites left their mark, no doubt, but they either found vacant land to occupy, or they displaced native inhabitants, or they were assimilated into the existing cultures.

Sons of Ham

Leaving the Japhethites, and turning our attention to the Hamites, we see much the same story - local infiltration, not global migration.

Genesis 10:6. The sons of Ham are: "Cush, and Mizraim, and Phut, and Canaan."

The descendants of Cush are identified with Eastern Mesopotamia, Arabia, Southern Asia and, perhaps erroneously, with Ethiopia. None of Phut's children are named in biblical history, but he appears to relate to the Libyans with some descendants pushing out as far as Mauritania.

The prophet Isaiah alludes to the local nature of the earliest migrations. "And it shall come to pass in that day, that the Lord shall set his hand again the second time to recover the remnant of his people, which shall be left, from Assyria, and from Egypt, and from Pathros, and from Cush and from Elam, and from Shinar, and from Hamath, and from the islands of the sea" (Isa. 11:11).

We can see how restricted the area was from Isaiah's vantage point. In the next verse, Isaiah speaks of the "outcasts of Israel" and the "dispersed of Judah" being brought "from the four corners of the earth" (Isa. 11:12). This is a specific prophecy either for the gathering of the Jews, and the founding of the nation Israel in 1948, or it could refer to the gathering unto judgment in the last days.

Genesis 10:8: "And Cush begat Nimrod: he began to be a mighty one in the earth."

Genesis 10:10. Nimrod is called the "mighty hunter." "And the beginning of his kingdom was Babel, and Erech, and Accad, and Calneh in the land of Shinar."

One of the monuments taken from excavations at Uruk is a stela made out of black granite depicting a bearded leader hunting lions:

> He uses a spear in one example, bow and arrow in the other, for he is represented twice. There is no inscription; no setting is indicated and there are no followers. The occasion of the hunt remains a mystery ... [3]

And similarly, the exploits of Nimrod remain a mystery. Unlike Asshur, no traces of Nimrod have yet been found in the annals of ancient history except for a city named Birs Nimrud.

Sargon of Agade

The middle of the third millennium was tumultuous in war-ravaged Sumer. Fortress cities fought for supremacy against each other with first one, and then another, vying for and gaining control, only to lose it again in yet another battle. Umma and Lagash were two of the cities most in opposition, and engaged one another in conflict frequently. Around 2450 BC, a monument was raised in Lagash heralding the triumph of King Eannatum over the Ummans.

At about 2375 BC, Lugalzaggesi of Umma exacted revenge and sacked the city of Lagash, burning their temples. [4] The heady aroma of victory was not to endure as the forces of Lugalzaggesi were crushed at Nippur by Sargon the Great, bringing to power a new ruler, a Semite, over all of Sumer and Accad.

Sargon's origins are surrounded in intrigue. Legend has it that a Semite woman found herself embarrassingly with child. After birth, baby Sargon was placed in a reed basket waterproofed with pitch, and set adrift Moses-style upon the headwaters of the Euphrates. The basket drifted downstream and was found by a Sumerian farmer irrigating his fields who later reared little Sargon in his home.

Sargon came from Nimrod's city of Accad, though some historians give Sargon credit for its founding. Starting his career as cupbearer to the king of Kish, Sargon launched an effective military campaign, and after crushing Lugalzaggisi's army, began his rule around 2371 BC. [5] A dynamic, conquering ruler, Sargon left his indelible mark.

> The Sumerian kingdom of Lugalzaggisi soon fell entirely into the hands of Sargon, whose full title proudly proclaimed him to be lord and master of the 'four quarters of the world,' namely Amurru to the west, Subartu to the north, Sumer and Accad to the south and Elam to the east. [6]

One of the key factors giving Sargon's rule such historical punctuation was that the inscriptions from his reign were set down in a
previously unrecorded language - a Semitic language. A shift to a
Semitic tongue after centuries of Sumerian inscriptions was a real
attention getter for archaeologists.

> In Mesopotamia, Sumerian was gradually replaced
> in official documents by Semitic Akkadian, though
> Sumerian was retained in the temples. [7]

Please note that two distinctly different cultures, Semite and
Sumerian, speaking unrelated languages, were together in the same
region at the same time. Not only were there different languages
in the world, but different languages were spoken in Southern
Mesopotamia prior to the tower of Babel incident, and even before
the flood, as both Sumer and Accad can be traced to about 4000
BC. Although many modern languages can be linked with Hebrew, there is no known language that can be traced to Sumerian.
Also note, the Sumerian language predates the Semitic.

> The Sumerian language is quite different in structure
> and vocabulary from any other known language of the
> ancient world, and attempts to derive the Sumerians from
> an original home in the Caucasus mountains, or from the
> Iranian plateau, on linguistic grounds have so far failed. [8]

A short testimony from Sargon has been pieced together from
two incomplete Neo-Assyrian tablets and one Neo-Babylonian fragment. It begins, "Sargon the mighty king, king of Agade, am I."
Sargon goes on to tell of his humble birth. Following are four lines
of the thirty-two line text:

> *And for four and [...] years I exercised kingship.*
> *The black-headed [people] I ruled, I gov[erned];*
> *Whatever king may come up after me*
> *   [...],*
> *Let him r[ule, let him govern] the black-headed*
> *   [peo]ple;...* [9]

Building on what we already have learned, these words from the Semite king imply that he was ruling over a people whose racial characteristics were different from his own people. Clearly, it refers to the Sumerians.

Dating the Flood

Historians place the beginning of Sargon's reign at 2371 BC, and that presents a challenge to the traditional date of the flood. Had the flood taken place at 2348 BC, according to the Ussher chronology, how could Sargon, a Semite, have been reigning 23 years before the rain, so to speak? The Bible does not name him as any of Shem's immediate kin, so Sargon must have been down the list on Shem's family tree.

Difficulties in reconciling biblical history with secular history, and archaeological dates with biblical dates, is nothing new. *The Ryrie Study Bible* offers one example of how perplexing this problem can be.

> **Date of the Exodus** Two principal views exist concerning the date of the Exodus: c. 1445-1440 B.C. during the reign of Amenhotep II (1450-1425) or c. 1290 B.C., during the reign of Raamses II (1299-1232). Scriptural evidence for the earlier date includes the statement of I Kings 6:1 that the Exodus occurred 480 years before the fourth year of Solomon's reign, thus placing it c. 1445. Further in Judg. 11:26, Jephthah (c. 1100 B.C.) declared that Israel had possessed the land of Palestine for 300 years, which would date the Exodus c. 1400 B.C.
>
> Objections to this earlier date include the following: (1) The Exodus could not have taken place until after 1300 because the city of Raamses was named after the pharaoh who was ruling at that time, and Raamses did not rule until 1299. However, if the Exodus was about 1290 and Moses was 80 at that time, and since the city was built before Moses' birth, the Exodus could not have been as

late as 1290 since there is no room for the 80 years of Moses' life between 1299 and 1290.

(2) It is said that the presence of strong opposition to the Israelites from the Edomites (Num. 20:20-21) was impossible before 1300 since the area of southern Transjordan was unoccupied from 1900-1300. Yet excavations in that area have uncovered objects and pottery dating as early as 1600.

(3) It is claimed that Hazor did not fall to the Israelites until 1300 B.C. However, Scripture states that it fell twice: first in the days of Joshua (Josh. 11:10-11) and later in the time of Deborah and Barak (Judg. 4:2, 23-24). Further there is evidence in one area of the excavated city of a destruction around 1400.

(4) The destruction of Lachish (Josh. 10:32-39) is said to have occurred 1230-1200 B.C., indicating a late date for the Exodus. But the book of Joshua does not claim that these cities were completely destroyed by Joshua (as Jericho was). Further, since the Stele of Pharaoh Merneptah represents the Hebrews as settled in Canaan when Merneptah's armies attacked them about 1230, the Exodus had to be somewhat earlier than 1290. [10]

Ryrie concludes with a lukewarm endorsement of the earlier date, 1440 BC, agreeing with Wood, Davis, Unger, and Archer. [11] Harrison, Wright, Kitchen, Albright and others prefer 1290 BC. A 150 year ambiguity in the date of the exodus from Egypt affects the estimated date for the flood since all estimates of dates have been done backwards from present to past.

To further confound the issue, Bible texts themselves disagree. Kraft brought out the problem of textual variations:

Here the figures vary considerably between those in the Hebrew Masoretic text (A.D. 600-900), from which a part of our English Bible is translated, and the figures in such early texts and versions as the Samaritan Pentateuch (originating before 200 B.C.) and the Greek Septuagint (originating about 250 B.C.). For example, in the Hebrew text the number of years between creation and the flood adds up to 1,656 years, whereas in the Greek it amounts to 2,242, and in the Samaritan only 1,307. [12]

Not only do the years from Adam to the flood vary from 1,307 to 1,656 to 2,242 between the three texts, the differences in years from the flood to Abraham are even more startling. The Masoretic text puts 292 years between the flood and Abraham's birth, whereas the Samaritan Pentateuch records 942! The latter figure is in complete agreement with the Septuagint except the Greek text lists another patriarch, Cainan, between Arphaxad and Shelah, adding another 130 years. Cainan is dutifully recorded in Luke 3:36, but missing in the received text from which a large part of our English language Bible is derived, adding weight and credibility to the oft-neglected Septuagint. [13]

Credit for the 4004 BC date goes to Archbishop Ussher and the received text. Ramm comments:

> The date of 4004 B.C. was tagged on to the Bible well before the founding of modern geological theory. This date goes back to the work of James Ussher (1581-1656), an Irishman and Archbishop of Armagh. [14]

Ramm goes on to lament that the "dates of Ussher have been almost canonized" due to their inclusion in our English language Bibles for centuries.

Raising ignorance to an art form, Lightfoot refined Ussher's work. Adam's creation, according to the famed Hebrew scholar from Cambridge, took place on October 23, 4004 BC at approximately 9:00 AM (forty-fifth meridian time, of course). [15] This led Brewster to remark:

> Closer than this, as a cautious scholar, the Vice-Chancellor of Cambridge University did not venture to commit himself. [16]

If we can learn from the many mistakes of others, it may do us well to exercise caution, and concentrate more on the sequence of events and their approximate dates of occurrence. Pertinent data can guide us toward that end. One bit of corroborating evidence reported in *The Cambridge Ancient History* relates to the date of the flood.

> This important landmark in the dawn of history was associated with the person of a Sumerian king named Ziusudra who was reigning at Shuruppak, precisely where a clean flood stratum has been found. [17]

Mallowan, from analysis of the flood deposits at Fara (Shuruppak), put the flood date at about 2900 BC. [18] According to the Watelin-Langdon chronology, the flood deposit at Kish was dated at 3000 BC. [19] The Early Dynastic Period in Mesopotamian history commences with the first post-flood rulers at Kish starting at 2900 BC, and ending at 2371 BC with the arrival of Sargon, who begins the Sargonid Period.

Between Ussher's estimated biblical date for the flood and Mallowan's archaeological date is some 550 years, not a particularly alarming disparity. If young-earth creationists can live with discrepancies in the billions of years, a mere 500 or 600 should not be all that disturbing. For those who prefer less ambiguity, though, there is another approach.

If we set the date of the Exodus to coincide with the reign of Raamses II, preferred by most Old Testament scholars today, and the years from the flood to Abraham is taken from the Septuagint, the quoted text of the New Testament authors, that would yield a date of 2978 BC, almost exactly the same as the archaeological date. If the Exodus took place in the reign of Amenhotep II, the biblical date for the flood would be 150 years sooner, still close to the archaeological date.

Calculating the Flood

Archbishop Ussher's Creation Date:	4004 BC
Years to the Flood (Masoretic Text)	- 1656
The Flood	2348 BC
Duration of Flood - 1 Year	- 1
The Flood Ends	2347 BC
Years to Abraham (Masoretic Text)	- 292
Abraham's Birth	2055 BC
Years Abraham to Exodus	- 615
The Exodus (Early Date)	1440 BC

The Exodus (Early Date - 1440 BC)	1440 BC
Years Abraham to Exodus	615
Abraham's Birth	2055 BC
Years Flood to Abraham (Septuagint)	1072
Duration of Flood - 1 Year	1
Date of Flood (Early Date)	**3128 BC**

The Exodus (Late Date - 1290 BC)	- 150
Date of Flood (Late Date)	**2978 BC**

The Asshur Expedition

Genesis 10:11: "Out of that land went forth Asshur, and builded Nineveh, and the city of Rehoboth, and Calah." Asshur began the Assyrian empire in the northeast corner of Upper Mesopotamia where the Tigris river runs from northwest to southeast. Here mounds of ruins can still be found today along both banks of the river.

The knowledge about some of the cities buried under
these mounds was never lost. That the mound of Nimrud
on the east bank, close to the point where the Greater Zab
flows into the Tigris, was the town of Kalakh mentioned
in Genesis 10:11 was told by the natives to a British representative of the East India Company who explored the
site in 1820. They even knew that the country to which
this town had once belonged was named 'al-Assur'. [20]

The data in Genesis 10:11, confirmed by archaeology and history, once again underscores the reliability of the biblical account.
Building cities, however, does not mean Asshur started from scratch
on virgin soil. Although the excavations at the city of Asshur only
hint at previous habitation, the underlying ruins beneath Nineveh
revealed levels of occupation that preceded the arrival of the
Assyrians by roughly 1,000 years. The artifacts recovered at
Nineveh were related to the pre-flood Ubaid or Halafian cultures,
not to the Semite or Sumerian peoples.

The comparison suggests that the early population of
Assyria was more or less identical with the pre-Sumerian
population of southern Babylonia; the absence of any remains corresponding to the so-called Early Dynastic period of Babylonia makes it clear, on the other hand, that
the Sumerians never occupied Assyria. [21]

Pottery and artifacts dating to the pre-flood period at Nineveh
is unmistakable evidence of a pre-existing populated site upon
which Asshur could build an Assyrian city, but the testimony runs
deeper then that. Cuneiform writing found at the site reveals the
city already was called "Ninua" before the Semites discovered
it. [22]

In the Land of Pharaohs

Genesis 10:13, 14. Mizraim and his sons are associated with Egypt. By no means does that signify they gave birth to the entire Egyptian populace. With the exception of the Philistines, who came from Casluhim with reinforcements from Caphtor, the rest of Mizraim's sons leave only sparse traces in various parts of Egypt. Pathrusim is associated with the island of Pathros where John was exiled.

Excavations in Egypt have uncovered the remains of a variant race of peoples who began moving in at the beginning of the dynastic period.

> Physically these peoples differed unmistakably from the predynastic Egyptians: whereas the latter were unusually small in stature and possessed long and narrow skulls (about 132 mm. in breadth), the newcomers were more massively built and their skulls (about 139 mm. in width) were appreciably broader than those of their predecessors. [23]

Although the origin of these immigrants is unclear from purely fossil findings, Semitic (or Hamitic) elements began to appear in the vocabulary of the Egyptian language, and it is entirely possible that these newcomers brought it with them. [24] Semitic (or Hamitic) traces appeared in Egypt after the first dynasty began with Narmer, who ruled about 3100 BC, and are not found prior to that date.

The existence of ancient peoples living in Egypt has been well established. The predynastic period in Egypt dates to 4000 BC. There is also evidence of population concentrations in the floodplain of the Nile dating as far back as 6000 to 5000 BC. [25] The building of the Aswan dam encouraged salvage programs in that area that uncovered "a virtually uninterrupted sequence of occupation from the Paleolithic to the Mesolithic in southern Egypt - that is, from about 30,000 to 10,000 B.C." [26] Not to belabor the point, but there is also evidence that the Nile valley was first explored as early as 700,000 years ago. [27]

It might do well to say here that Bible historians in the past have made attempts to join up disparate populations and find links to Noah on the premise that all the world's populations are descendants. Secular anthropologists, archaeologists, and historians are not so restricted. As a result, the best matches between biblical and secular history have come from those investigators who have not allowed their work to suffer from prejudiced assumptions.

The Cursed Canaanites

Genesis 10:15-18. From Noah's grandson, Canaan, came Sidon and Heth, followed by the Jebusites, Amorites, Girgasites, Hivites, Arkites, Sinites, Arvadites, Zemarites, and the Hamathites.

Sidonians dwelt at the "northern borders of Canaan or Phoenicia." [28] The Hittites are the sons of Heth, and initially occupied a stretch of land south of the Black Sea before they began their conquering ways. Jebusites inhabited Jerusalem. Amorites remained closely associated with the Canaanites, and ranged from "the mountains of Judah and beyond the Jordan in the time of Moses." [29]

The Arkites found their space in the south of Lebanon; also the home of choice for the Sinites, though no one seems to know exactly where. The Arvadites took up residence on a "small rocky island of Arados to the north of Tripolis." [30] Zemerites were the "inhabitants of Simyra in Eleutherus." [31] The town of Hamath, located about 115 miles north of Damascus, was founded by the Hamathites.

Genesis 10:19: "And the border of the Canaanites was from Sidon, as thou comest to Gerar, unto Sodom, and Gomorrah, and Admah, and Zeboim, even unto Lasha."

The Canaanites took the southeastern shore of the Mediterranean Sea. In 1977, the Canaanite city of Ebla was brought to light at Tell Mardikh in Syria. Dating to the Chalcolithic period, Ebla appears to have been a major trading partner with Mari and Uruk. The clay tablets excavated from Ebla revealed a cuneiform style of writing similar to that found at Shuruppak and Abu Salabikh, dating to the same period.

Canaanites, descendants of Ham, spoke a Semitic dialect. (One might say a Hamitic dialect.) The language "belongs to an archaic phase of the Northwest Semitic languages, which include Ugaritic of 1400-1200 B.C., Phoenician, and Biblical Hebrew." [32] In other words, the sons of Noah ended up speaking virtually the same language meaning there was never a scrambling of tongues in the sense that their basic language might have been permanently altered.

Today there are roughly 5,000 languages in the world of which between 50 and 200 are unrelated stocks. By contrast, Accadian, Canaanite, and Amorite languages, as well as later Hebrew, are quite similar and closely related. The building of the tower and the confusion that brought construction to a halt was an event affecting only those listed in that same chapter, the descendants of Arphaxad, son of Shem.

Other relatives, nationalities, and maybe slaves too, could have joined in, but we can only speculate on who they might have been; Hamites descended from Nimrod, maybe; local Sumerians, perhaps. But the idea that the population of the entire world was concentrated at Babel, and that all the world's languages precipitated from the confusion of tongues, is ill conceived, and has inadequate biblical support.

Sons of Shem

In contrast to the migrations of the Japhethites who actually crossed a body of water, and the Hamites, some of whom emigrated to northern Africa, the Semites were veritable home bodies.

<u>Genesis 10:22</u>. The children of Shem are: Elam, Asshur, Arphaxad, Lud, and Aram. Scholars have placed Elam at the head of the Elamites, a parallel culture to the Sumerians located to the east on the Persian plateau. The difficulty is that Proto-Elamite artifacts have been found dating as far back as 3200 B.C. [33] From excavations at Susa, the capitol city of Elam, archaeologists have uncovered their written tablets. The extinct Elamite language is unrelated to Semitic, Sumerian, or any other known languages.

Any relationship between Elam and the Elamite nation is problematical. Did Elam himself and some of his heirs find a way to win the hearts and minds of the fierce Persian predecessors? About 2700 BC, the Elamites conquered Sumer for a brief time. [34] Did they take on Semite advisors who eventually attained the reigns of power? The Achaemenids, precursors to the Persians, left inscriptions in ancient Elamite, but eventually came to speak Aramaic. [35] Where and when did they acquire a Semitic tongue? Or was "Elamite," the Accadian word for "highlander," used to describe those who came from that mountainous area? Well, we don't know. For the time being, the derivation of the early Elamite empire remains as much a mystery as the origin of the Sumerians.

A possible solution is that the Semites may have named some of their children after geographical areas or towns, whereas Bible interpreters typically have thought the Semites were founding fathers. A clue can be found in Genesis 11:26-32. Terah named his third son Haran, and Terah died in the city of Haran. Since the city must have been in existence before Terah's son, this implies his son was named for the city.

In Genesis 2:11, Havilah is a land where there is gold, bdellium and onyx. The Mesopotamian valley is devoid of precious stones, copper, or gold. These items had to be bartered in exchange for cereal grains the Semites and Sumerians harvested in abundance from the rich alluvial soil. From Gowlett:

> Raw materials other than clay were so utterly lacking on
> the Mesopotamian plains that almost everything needed
> to be imported. [36]

Havilah, the son of Joktan, may have been named for a land known for its gold. Likewise, Elam may have been named after the land of the Persians, perhaps for their fierceness or some other quality associated with that nearby culture.

Genesis 10:25: "And unto Eber were born two sons: the name of one was Peleg ..." Peleg's name means "to divide," and was so named because "in his days was the earth divided." This has been taken by some as possibly pointing to the dispersion at Babel, but there is another explanation. Kings of nation states in Sumer fought continuously with one another. A war-torn land is a divided land with divided peoples.

From Lud came the *Lydians* who may have remained in the same general area as the Assyrians, though Bush places them in Ethiopia. [37] The *Aramaeans*, founded by Aram, situated themselves in various parts of Syria and Mesopotamia, and from them the Chaldeans descended. The children of Aram - Uz, Hul, Gether, and Mash - are all to be found in close proximity to the same area settled by their father. Joktan is considered to be the head of the primitive Arabian tribes; his sons can be traced largely to places and districts in Arabia. [38]

Races Not Traceable to the Sons of Noah

Writing in 1888, Alexander Winchell found the wanderings of Noah's sons and all their kin "does not embrace more than one-fifteenth of the territory which we now find populated by man." [39] Bristling at the theological establishment, Winchell made a caustic comment on Genesis 10:

> Was this an attempt to explain the origin of all the nations of the world? Does this genealogical map imply that the regions beyond its limits were then unoccupied by human beings? Does it mean that the various tribes and nations which are now spread over the earth have arisen from the wider dispersion of the sons of Noah? Have the black tribes of Africa and Australia and Melanesia, and the brown nations of Asia and America and Polynesia, been produced from the posterity of Noah during the interval which separates us from the flood? Yes, says the catechism, which under cover of religious instruction, assumes to indoctrinate our children in ethnological science. Yes, yes, says the commentator, who experiences no difficulty in swallowing the exegetical and indigestible crudities which have been the heirlooms of the church for two thousand years. Yes, yes, yes, exclaims,

> too unanimously, the modern teacher of "divine truth" all
> unconscious that the science of ethnology has made vis-
> ible advances since Jerusalem was the center of the world.
>   To all these questions I reply in the negative.[40]

In addition to the relatively recent migrations presenting an insurmountable obstacle, the almost entirely local confines of the Hamites precludes them from being the forerunners to the black Africans, for one example, as has been advocated by many Bible interpreters.

Postulating the Caucasians as emanating from Japheth is equally fraught with difficulty. And no one yet has proposed any descendant of Noah's three sons who could have been ancestral to the native Americans, whose fossil remains date to as early as 11,600 years ago, [41] and who must have crossed the Bering Strait during the last ice age, 40,000 to 12,000 years ago when there was a land bridge. [42]

The Chinese civilization also has traces into the distant past. Recent archaeological endeavors in China have confirmed the antiquity of the Chinese culture:

> Until recently, the Chinese Neolithic has given the im-
> pression of appearing fully fledged at about 5000 BC, but
> earlier stages of the story are now becoming apparent.
> Hsienjentung, a site in Kiangsi, appears to be older than
> 6800 BC ... [43]

The Australian Aborigines settled the "land down under" at least 38,000 years ago. They, among others, cannot be seen to have roots leading to Noah.

The Ice Man Cometh

The mummified body of a Tyrolean mountaineer, dubbed the "Ice Man," was first spotted September 19, 1991, in the melting Similaun glacier high in the Alps. Radio-carbon dating placed this newly-famous, tattooed wanderer in the Neolithic Period, between 5,100 to 5,300 years ago. A flint-bladed dagger and copper axe found along with his body corroborated this astonishing date. [44] A report by *Time* went on to say:

> ... it is clear that he had well-formed facial features that would not draw stares from contemporary Tyroleans. Says South Tyrolean archaeologist Hans Notdrfter: "He looks like one of our well-tanned ancestors." [45]

An adjacent article in that same issue of *Time* included data bits about the world as it was in the day of the Iceman:

> ... by about 4000 B.C. there were an estimated 86.5 million people on earth, about eight times as many as there had been 2,000 years earlier. [46]

Put in perspective, the world had over 80 million people in residence a thousand years before the flood, who seem to be ancestral to modern-day human beings. And about ten million people were scattered around the globe a thousand years before Adam!

Later DNA testing confirmed that the Ice Man's mitochondrial DNA type fits into the genetic variation of contemporary Europeans, and is "most closely related to mitochondrial types determined from central and northern European populations" [47] - particularly Germany, Denmark, and Iceland. Gene sequencing showed that within the European gene pool, the Ice Man appears to be "more closely related to contemporary Alpine and northern populations." [48] Considering the 3100 to 3300 BC date, we have hard evidence of a European mountain man no different racially from present-day indigenous populations, who lived at the time of Noah, before the flood, and who died before Shem, Ham, and Japheth were even born!

The sons of Noah may have founded nations, but they also found nations whose history, customs, and traditions were unlike their own, who spoke different languages, and whose skin color was lighter in the north and darker to the south.

NOTES

1 George Bush, *Genesis* (New York: Iveson, Phinney & Co., 1860), 169.
2 C. F. Keil and F. Delitzsch, *Commentary On The Old Testament* (Grand Rapids: William B. Eerdmans Publishing Company, 1986), 164.
3 I. E. S. Edwards, C. J. Gadd and N. G. L. Hammond, eds., *The Cambridge Ancient History* Vol. 1, Part 2 (Cambridge: Cambridge University Press, 1971), 80.
4 George Constable, ed., *The Age of God-Kings: Time Frame 3000-1500 BC* (Alexandria: Time-Life Books, 1987), 32.
5 C. C. Lamberg-Karlovsky and Jeremy A. Sabloff, *Ancient Civilizations: The Near East And Mesoamerica*, (Menlo Park: The Benjamin/Cummings Publishing Company, Inc., 1979), 163.
6 Andre Parrot, *Sumer* (France: Thames and Hudson, 1960), 170.
7 S. G. F. Brandon, ed., *Ancient Empires: Milestones of History* (New York: Newsweek Books, 1970), 21.
8 Ibid., 20.
9 James B. Pritchard, *Ancient Near Eastern Texts Relating to the Old Testament* (Princeton: Princeton University Press, 1955), 119.
10 Charles C. Ryrie, *The Ryrie Study Bible* (Chicago: The Moody Bible Institute, 1978), 91.
11 Gleason L. Archer, *A Survey of Old Testament Introduction* (Chicago: Moody Press, 1974), 229-241.
12 Charles F. Kraft, *Genesis: Beginnings of the Biblical Drama* (New York: Woman's Division of the Christian Service Board of Missions, The Methodist Church, 1964), 74-75.
13 James A. Borland, "Did People Live to Be Hundreds of Years Old Before the Flood?" *The Genesis Debate*, ed. Ronald F. Youngblood (Grand Rapids: Baker Book House, 1990), 169.
14 Bernard Ramm, *The Christian View of Science and Scripture* (Grand Rapids: William B. Eerdmans Publishing Co., 1954), 121.
15 Edwin Tenney Brewster, *Creation: A History of Non-Evolutionary Theories* (Indianapolis: The Bobbs-Merrill Company, 1927), 108-109.
16 Ibid., 109.
17 Edwards, Gadd and Hammond, eds., *The Cambridge Ancient History* Vol. 1, Part 2, 244.
18 M. E. L. Mallowan, "Noah's Flood Reconsidered," *Iraq* Vol. 26, Part 2 (Autumn, 1964), 81.
19 Andre Parrot, *The Flood and Noah's Ark* (New York: Philosophical Library, 1953), 49.

20 Edwards, Gadd and Hammond, eds., *The Cambridge Ancient History* Vol. 1, Part 2, 729.

21 Ibid., 730.

22 Ibid., 730.

23 Ibid., 40.

24 Ibid., 41.

25 Lamberg-Karlovsky and Sabloff, *Ancient Civilizations: The Near East And Mesoamerica*, 130.

26 Ibid., 126.

27 From a pamphlet *The First Egyptians* published by the McKissick Museum and Earth Sciences and Resources Institute of The University of South Carolina.

28 Bush, *Genesis*, 169.

29 Keil and Delitzsch, *Commentary On The Old Testament*, 169.

30 Ibid., 169.

31 Ibid., 169.

32 Giovanni Pettinato, *The Archives of Ebla* (New York: Doubleday & Company, Inc., 1981), 7-10.

33 Lamberg-Karlovsky and Sabloff, *Ancient Civilizations: The Near East And Mesoamerica*, 152.

34 Ibid., 181.

35 Jim Hicks, *The Persians* (Alexandria, Time-Life Books, 1975), 9-12.

36 John Gowlett, *Ascent To Civilization* (New York: Alfred A. Knopf, Inc., 1984), 180.

37 Bush, *Genesis*, 169.

38 Keil and Delitzsch, *Commentary On The Old Testament*, 170.

39 Alexander Winchell, *Preadamites; or a Demonstration of the Existence of Men Before Adam* (Chicago: S. C. Griggs and Company, 1888), 89.

40 Ibid., 89.

41 "Antiquity of oldest American confirmed," *Science News* vol. 142, No. 20 (Nov. 14, 1992), 334.

42 Roger Lewin, *In the Age of Mankind* (U. S.: Smithsonian Institution, 1988), 159.

43 Gowlett, *Ascent To Civilization*, 166.

44 Leon Jaroff, "Iceman," *Time* (October 26, 1992), 62-63.

45 Ibid., 64.

46 Ibid., 67.
47 Oliva Handt et al., "Mitochondrial Genetic Analyses of the Tyrolean Ice Man," *Science* (17 July 1994), 1775.
48 Ibid., 1778.

Chapter 17

THE TOWER OF BABEL: LESS CONFUSING

After the dispersing, settling, and, probably, conquering by the tribal descendants of Noah's three sons, a landmark incident took place in the plain of Shinar that caused and still causes confusion - at the tower of Babel.

Genesis 11:1: "The whole earth was of one language, and of one speech."

It is not difficult to see how Bible interpreters have been as baffled as the tower builders. The true confusion of tongues, surpassing the incident at Babel, is the translation of Hebrew into English. Yet again, 'erets is translated "earth," although in the next verse the same word is rendered correctly as the "land" of Shinar.

Any other duly authorized word such as land, district, region, or territory would have suited the occasion adequately. It's the word "earth" in this and verses discussed previously that has become the launching pad propelling Bible exegesis into outer space. It is long past time to abandon this circular, non-productive orbit of misinterpretation, and get back down to - not "earth" - but land.

Burnt Brick Set in Bitumen

Genesis 11:2-3: "And it came to pass, as they journeyed from the east, that they found a plain in the land of Shinar; and they dwelt there.

And they said one to another, Go to, let us make brick, and burn them thoroughly. And they had brick for stone, and slime [asphalt, bitumen] had they for mortar."

The children of Arphaxad, descendants of Shem, named in the same chapter of Genesis, are the likely builders of the tower at a time in history when tower building was all the rage.

> In Mesopotamia, the temples of the predynastic period developed into grandiose monuments which dominated not only the cities they were meant to serve, but the whole of the valley floor. It has even been suggested that the ziggurats, the stepped mounds which supported the sacred shrines, were intended simply as artificial mountains. Though their design showed high skill, technically they were of the simplest: a mudbrick core encased in a weatherproof skin of burnt brick set in bitumen. [1]

Although the Egyptian pyramids were constructed of cut stone, the Mesopotamian ziggurats were, in fact, constructed with mudbricks and burnt as stated in Genesis 11:3. The bitumen used as a weathertight casing comes to us as "slime" in the King James version though the New American Standard uses "tar."

> The evolution of the temple complex is well illustrated by the Anu Temple in Uruk. Six temples were constructed, one above the other.... after five hundred years of rebuilding, a monumental brick platform rose 16 meters above the community. [2]

Not only the Semites, but the Sumerians too, were adept at building ziggurats. In addition to the site at Uruk, the Sumerians built temple monuments at Nippur, Lagash, Kish, and Ur. Even smaller population centers to the north were building their own. Many of the true ziggurats were built upon old temple complexes about the time of the Third Dynasty at Ur (2112-2000 BC); some, perhaps, were constructed a little earlier.

A Ziggurat Building Contest

A picture emerges of the region as a land of city-state kingdoms laced with an intricate canal network for farming and facilitating trade. Although trading between cities was an integral part of the culture, fierce competition was quite evident, and sacking neighboring cities was all too common.

Growing populations required more water, and as northern cities extended their network of canals to irrigate additional fields, they deprived their southern neighbors of life-giving water. In anger, the southern cities would wage war on their northern neighbors and wreak havoc on their irrigation systems. Cities would recover in time and exact their revenge, starting another cycle.

In their literature, these continual battles with ensuing changes of kingship are interspersed with the names of the kings and the years they reigned. Leaving out those details, following is a brief sequence:

> *The weapons of Kish were overthrown;*
> >   *its kingdom passed to Opis.* (Kings omitted)
> *The arms of Opis were overthrown;*
> >   *its kingdom passed to Kish.* (Kings omitted)
> *The arms of Kish were overthrown;*
> >   *its kingdom passed to Erech.* (Kings omitted)
> *The arms of Erech were overthrown;*
> >   *its kingdom passed to Agade, etc.* [3]

Individual cities, united for waging war, were also sufficiently organized for civic building projects. At each cult center, a simple temple mound was erected dedicated to a particular god. Although these mounds can be traced to as early as 3000 BC, by the end of the third millennium they were reaching immense proportions.

The Ashmolean Prism contains a liturgy to the temple at Kes, presumed to have been in the proximity of Erech and Shuruppak. Numerous lines end with "attaining unto heaven":

> Oh temple whose design in heaven and earth has been
> planned, thou are possessed of pure decrees. Temple
> erected in the Land where stand the chapels of the gods.
> Mountain house, radiant with abundance and festivity. [4]

A kind of ziggurat contest ensued as cities added mudbrick platforms on top of older temple complexes topped with granite, sandstone, and marble temple enclosures. It became a point of honor and pride to outdo neighboring cities, and of course, this demonstrated love and devotion to their deity.

Genesis 11:4: "And they said, Go to, let us build us a city and a tower, whose top may reach unto heaven; and let us make us a name, lest we be scattered abroad upon the face of the whole earth."

Gudea, king of Lagash, sings the praises of the temple he built to worship Eninnu constructed from burnt bricks, stone, and "bitumen from wells," and "bitumen from bitumen lake." [5] In addition to the similarity in building materials, note also the similarity in attitude and aspirations with the builders at Babel. In Gudea's words, the temple was "an object of admiration to the eyes of the gods."

A sense of this pride and competitive spirit that was emblematic of virtually every city in Mesopotamia can be seen in the following liturgy from king Gudea:

> The bright crown of the temple rested upon it and as the
> lapis-lazuli mountain of heaven and earth rose from the
> earth. The pavement of the terrace of the great temple he
> laid; as a pure vessel on which honey and wine are poured
> it was open to heaven. The shrine with a couch which he
> built like a perfect mountain, as the holy stone vessel of
> the deep it rose. On account of the great name which he
> had made for himself he was received among the gods
> into their assembly. [6]

Notice the commonality between "the great name which he had made for himself" in the liturgy with "let us make us a name" in Genesis 11:4. The Semite builders at Babel were not about to be outdone. After all, they had the one true God, and the other cities were honoring impostors. They would build a tower of such pro-portions that it would show up the others and prove to them whose God reigned supreme. You can imagine how happy God would be to have a mound of mudbricks 10 feet higher than one dedicated to the Accadian sun god, Shamash, or the Sumerian moon god, Nanna. This misguided endeavor was not edifying to say the least.

Genesis 11:8: "So the Lord scattered them abroad from thence upon the face of all the earth: and they left off to build the city."

Many Towers

The traditional interpretation of the flood and the dispersion at Babel has been that the total population of the entire world was confined to the land of Shinar in the post-flood era. They were all related, according to tradition, all spoke a common language, and they became engrossed in building the tower at Babel. The Lord confounded them, and off they went in all directions muttering Aztec, Mandarin, Swahili, and the like. They crossed oceans and reached far distant continents. They made changes in skin color too, presumably, and developed morphological adaptations, as they went along.

This interpretation has perpetuated in spite of the extra-biblical evidence available all along that nullifies this interpretation. All those Bible apologists had to do was count the mud brick ziggurats in Mesopotamia. Any number that exceeds one kills that explana-tion. Five or more ziggurats should seal the lid on the coffin. In fact, The Atlas of Mesopotamia locates over thirty ziggurats and temple mounds in the region including Persia. [7] Parrot identified thirty-three towers in twenty-seven different cities, adding "it was sometimes possible for one city to have several ziggurats." [8]

Had the entire earth been devoid of humanity except for Noah's tribes clustered together in the land of Shinar where the tower of Babel was built, what would explain the additional towers? The other ziggurats at various sites had to be constructed either before, at the same time, or after the tower of Babel. If the other monuments were constructed before Babel, it would mean that Noah's descendants had already begun to spread out and settle in widely separated communities, precluding them from all being at one place, which was the case according to Genesis 10.

If constructed all at the same time, however, that would infer multiple Babels and simultaneous dispersions at all locations. Who would like to argue for that? But if the other ziggurats were built after Babel, it would mean that after the Lord taught them a lesson and sent them packing, they gathered together and built more ziggurats all over the place like nothing had happened!

What we find is that building ziggurats was simply in vogue in those days. The tower of Babel was one among a number of Mesopotamian worship centers. Though we may never know with complete certainty which one it was, the mound at Babylon is a prime candidate. The following is part of *Enuma Elish*:

> *"Now O lord, thou who hast caused our deliverance,*
> *What shall be our homage to thee?*
> *Let us build a shrine whose name shall be called*
> *'Lo, a chamber for our nightly rest': let us repose in it!*
> *Let us build a throne, a recess for his abode!*
> *On the day that we arrive we shall repose in it."*
> *When Marduk heard this,*
> *Brightly glowed his features, like the day:*
> *Like that of lofty Babylon, whose building you have*
> *   requested,*
> *Let its brickwork be fashioned. You shall name it 'The*
> *   Sanctuary.'"*
> *The Anunnaki applied the implement;*
> *For one whole year they molded bricks.*
> *When the second year arrived,*
> *They raise high the head of Esagila equaling Apsu.* 9
> *Having built a stage-tower as high as Apsu,*
> *They set up in it an abode for Marduk, Enlil, (and) Ea*
> *In their presence he adorned (it) in grandeur.*
> *To the base of Esharra its horns look down.*

After they had achieved the building of Esagila,
The Anunnaki themselves erected their shrines.
[...] all of them gathered
[...] they had built as his dwelling.
The gods, his fathers, at his banquet he seated:
"This is Babylon, the place that is your home!" 10

Clearly the building of the tower and the confusion of tongues at Babel loomed large to the participants, but the tower itself was one among many. It was not the biggest, and cannot be identified as the first or the last.

The Tower Restored and Rebuilt

The rebuilding of destroyed temples was carried out long after the heyday of Sumer and Accad. The ziggurat at Babylon was restored by Nabopolassar, the founder of the Neo-Babylonian dynasty, about 625 to 605 BC. These are his words:

> The lord Marduk commanded me concerning Etemenanki, the staged tower of Babylon, which before my time had become dilapidated and ruinous, that I should make its foundations secure in the bosom of the nether world, and make its summit like the heavens. 11

His firstborn son, Nebuchadnezzar, continued in the efforts started by his father, carrying out building the tower at Babylon until 562 BC. When finished, a seven stage structure and its temple complex reached nearly 300 feet in height. 12

Herodotus visited Babylon about 460 BC and gave this report:

> In the midst of the temple a solid tower was constructed, one stadium in length and one stadium in width. Upon this tower stood another, and again upon this another, and so on, making eight towers in all, one upon another.

All eight towers can be climbed by means of a spiral staircase which runs round the outside. About halfway up there are seats where those who make the ascent can sit and rest. In the topmost tower there is a great temple, and in the temple is a golden table. No idol stands there. No one spends the night there save a woman of that country, designated by the god himself, so I was told by the Chaldeans, who are the priests of that divinity. [13]

Sons of Arphaxad

Genesis 11:10: "These are the generations of Shem: Shem was an hundred years old, and begat Arphaxad ..."

Descendants of Shem and Arphaxad must have experienced those places of worship, constructed by their Sumerian neighbors, devoted to pagan gods, and decided to follow suit, building a monument of their own. Polytheistic worshippers erecting temples to pagan gods was tolerated due to their ignorance, but God's chosen were expected to exercise better judgment.

The Lord was not pleased with this enterprise, and put an end to their foolishness by confusing their speech. This was described by Oracles:

> And now all intercourse,
> By some occult and overruling power,
> Ceased among men: by utterance they strove
> Perplexed and anxious to disclose their mind;
> But their lip failed them; and in lieu of words
> Produced a painful babbling sound ... [14]

Then this particular band of Semites, maybe mixed with Sumerians too, dispersed. And judging from their writings, some of which pre-date the time of Babel, there was no permanent alteration in their basic languages.

Ziggurat Design

Although a tripartite design was common to many ziggurats, distinctive designs and shapes reflected regional differences. Unger identified three different types which he labeled as a rectangular type, a square type, and a combined type. Parrot added a fourth, which he called "a temple on a high terrace." 15

Typically, the Sumerians used a rectangular base with ramps for access. Examples of these were found in the south at Ur, Uruk, and Nippur, although triple staircases were added at Uruk and Nippur. 16 The Assyrians in the north preferred a square foundation with staircases, such as were found at Asshur and Nimrud. At Eridu, the style was the same as the Assyrian. The square foundation, emblematic of Semite construction, could signify a Semitic influence on the ancient city of Eridu.

The temple at Nineveh, built by the Assyrians, is over 260 miles from Babylon, farther than New York City is from Washington, D.C. Just as New Yorkers took no part in building the Washington monument, so too it is unlikely that Ninevites offered any assistance to build a tower in the plain of Shinar where Babylon is located, especially when they had a tower of their own to build.

There was considerable distance between many of the cities that contained Semite populations, cities that were founded before the building of the tower of Babel came to a halt. Even Shem's descendants could not all have been at Babel when the confusion of tongues occurred; not to mention the neighboring Sumerians, Elamites, Gutians, or Egyptians; or, for that matter, distant, primitive cultures concentrated in areas of the world known today as Mexico, Denmark, Thailand, and Japan. 17

Purpose for the Ziggurats

The impetus for building the first temple mounds is unknown. Were they platforms for saving lives in the event of floods? Could they have been a means of defense? When attacked, they could scamper up high platforms where they might hold the enemy at bay, much as medieval forts were utilized from where defenders could throw things down upon the attacking enemy.

The temples were also places of worship, where townspeople could honor their particular deity and offer sacrifices. Yet, none of these purposes explain why the temple structures began to become skyscrapers. From 3000 BC to about 2100 BC, more modest temple complexes in cities all over the region blossomed into imposing ziggurats.

The objective in constructing massive mudbrick structures is hinted at in the names chosen. "House of the mountain of the universe" stood at Asshur. "House of the seven guides of heaven and earth" was located at Borsippa. The "House of the king counsellor of equity" was at Ur, and the "Lofty house of Zababa and Innina whose head is as high as the heavens" was built at Nippur. Larsa had the "House of the link between heaven and earth," and at Babylon was the "House of the foundation of heaven and earth." 18

The Hebrew word *balal* means to confound or mix, and from Babel our English word "babble" is derived, defined as, "to utter meaningless or unintelligible sounds." 19 These definitions followed the event; however, the name "Babel" was not chosen in anticipation of confusion. The origin of "Babel" appears to be rooted in the Accadian word, *bab-ilu*, or "Bab-El," meaning, "gate of God." Considering Babylon as the likely location for the tower of Babel, then the name literally was "Gate of God" of the "House of the foundation of heaven and earth."

Between Beersheba and Haran, Jacob dreamed of a ladder, "set up on the earth, and the top of it reached to heaven: and behold the angels of God ascending and descending on it" (Gen. 28:12). He proclaimed that place "Beth-El," "the house of God, and this is the gate of heaven" (Gen. 28:17).

Did Jacob's grandfather Abraham describe to his son and grandson the tower at Babylon, or the one at Ur, in such vivid detail that it became a shadow-in-the-mist in Jacob's dream?

Come Down, Lord

Unlike our modern churches and tabernacles, these temples had a greater significance than just being a place to come and worship. They were intended as the "house" or dwelling place of their deity. The closer to heaven the towers could reach, the nearer the worshippers could approach, and it was a two way street; the deity could descend to the people more easily. By means of the tower itself, perhaps, they could invoke their god to come down. Parrot touches on the ziggurat as a "link."

> Thus the *ziggurat* appears to me to be a *bond of union*, whose purpose was to assure communication between earth and heaven. Even when this idea is not actually clearly expressed, it is nevertheless implicitly suggested; for what is the 'mountain' but a giant step-ladder by means of which a man may ascend as near as possible to the sky? Not only in order to touch it, but also, and especially, to approach nearer to the deity whom he seeks, and whose descent towards mankind he wishes at the same time to facilitate. [20]

Was this in the minds of God's chosen people, His faithful remnant, those in the line of Shem and Arphaxad? Perhaps, if the sanctuary was high enough, and if the sacrifices were appealing or in sufficient abundance, God Himself might be enticed to come down and dwell among His people. Isaiah voiced that appeal, "Oh that thou wouldest rend the heavens, that thou wouldest come down, that the mountains might flow down at thy presence ..." (Isa. 64:1).

Ultimately they were successful. God did come down, but not then, not striding down the steps of a Mesopotamian ziggurat. God came down over two thousand years later, in Bethlehem, in the land of Judea.

Historical Footnotes

Semites from the north migrated gradually into Sumer adapting to the local life style as they went along. From Hawkes:

> Semites, who had adopted Sumerian culture and adapted cuneiform to the writing of their own language, are known as Akkadians. The mingling of the two peoples and their traditions produced a vigorous civilization, but it was not long before the Semites were to become the dominant partners. [21]

The Sargonid period, also called the Accadian period of Mesopotamian history, lasted about 150 years and saw five kings come to power. The great Sargon ruled from 2371 to 2316 BC. His grandson, Naram-Sin, was the fourth in line of succession, and reigned from 2291 to 2255 BC. All of the city-states of Mesopotamia came under his rule; even the Elamite dynasty in Persia was subdued. [22]

The Semitic Accadians found out what subsequent empires throughout history have learned to their chagrin. Power is easier to grasp than maintain. Continual raids from without and revolts from within weakened the Accadian hegemony until the nomadic Gutians toppled the Sargonid dynasty around 2200 BC.

A period of turmoil followed the Gutian victory.

> The arrival of the Guti and their conquest of the Akkadians around 2200 B.C. ushered in a period of anarchy. The Sumerian King List asks "Who was king? Who was not king?", recording twenty-one kings in a period of ninety-one years. [23]

Utu-Legal became king of Uruk in about 2120 BC. His empire lasted seven years, and he suffered defeat at the hands of Ur-Nammu, governor of Ur. Subsequently, Ur-Nammu was crowned "King of Ur," and further took on the title of "King of Sumer and Agade." He began the Third Dynasty of Ur lasting from 2112 to nearly 2000 BC. [24]

It was during the reign of Ur-Nammu that the ziggurat at Ur was erected for the moon-god Nanna. [25] This massive temple monument dominated the landscape in the ancient city. Minority Semites may have been conscripts for this undertaking, necessitating a departure for any unwilling to endure another ziggurat project, honoring a pagan god to boot.

The Destruction of Ur

During the bloody reign of Naram-Sin, independence-minded cities were forced under submission to Accad. To guard against future uprisings, Naram-Sin ordered the fortress walls of Ur brought down. By complying, Ur became dependant on the military strength of the Semite king, and vulnerable to attack. Gutians and Elamites attacked the city and destroyed it about 2000 BC, slaughtering and enslaving nearly half a million Sumerians.

A Sumerian scribe set down what is now called "Lamentations Over the Destruction of Ur." These are a few lines:

> *On its walls they lay prostrate. The people groan.*
> *In its lofty gates where they were wont to promenade*
>   *dead bodies were lying about;*
> *In its boulevards where the feasts were celebrated*
>   *they were viciously attacked.*
> *In all its streets where they were wont to promenade*
>   *dead bodies were lying about;*
> *In its places where the festivities of the land took*
>   *place the people were ruthlessly laid low.* [26]

The scribe further lamented over the temple:

> *The lofty unapproachable mountain, Ekissirgal-*
> *Its righteous house by large axes is devoured ...* [27]

Naming the Gutians and Elamites as defilers of the temple, the scribe spat out his hatred against the "destroyers" who "made of it thirty shekels." [28]

To the Sumerians, "thirty shekels" signified degraded value, something of great value treated as if it had little value. What tragic irony that a Sumerian scribe would use a term of description also found in the Bible. The life of a slave was set at "thirty shekels" in Exodus 21:32; and in Matthew 26:15, a traitor named Judas was paid in similar measure for the life of a King.

The End of Sumer

Sumerian political authority over the region ended with the capture of Ibi-Sin, the last of the third dynasty of Ur. For two hundred years thereafter, Mesopotamia struggled with small, protective, city-state kingdoms, such as Assur and Eshnunna in the north, and Isin and Larsa in Sumer. [29] As for the Sumerian people:

> In matters of culture and religion, however, they continued to play a leading role for many centuries, while as an ethnic group they were slowly absorbed into their Semitic environment. We do not know when this process of absorption was completed. [30]

Sometime after the scattering at Babel, Abram's father, Terah, traveled to Ur.

Genesis 11:26-28. Terah was 70 years old, and had three sons. Terah's third son, Haran, died where he was born, "in Ur of the Chaldees."

Genesis 11:31, 32. Abram (later called Abraham) and his wife, Sarai (later named Sarah), journeyed with Terah from the city of Ur, either before, or possibly at the time of its destruction, toward a land of potential peace and quiet. Their travel plans were to go by way of Haran to the land of Canaan, though Terah stayed in Haran and died there.

It could have been rampant polytheism, part of the Sumerian-Semite culture at the time, that prompted Terah to send his family to a land settled by distant kin. Perhaps it was political turmoil or famine that prompted his journey, or it could have been something a few of us can relate to - war. Terah and his entourage, including Abraham, may have been refugees.

The last king of Larsa, Rim-Sin, was defeated by the great Babylonian king Hammurabi, bringing an end to the domination of the region by the Semitic Amorite kings. The date was shortly after 1800 BC, the new kingdom was now Babylonia, the peoples emerging from the region would come to be known only as Semites, and Sumer was undone.

NOTES

[1] Jacquetta Hawkes, *The Atlas Of Early Man* (New York: St. Martin's Press, 1976), 102.

[2] C. C. Lamberg-Karlovsky and Jeremy A. Sabloff, *Ancient Civilizations: The Near East And Mesoamerica* (Menlo Park: The Benjamin/ Cummings Publishing Company, Inc., 1979), 147.

[3] George A. Barton, *The Royal Inscriptions of Sumer and Akkad* (New Haven: Yale University Press, 1929), 351.

[4] Stephen Langdon, *Sumerian Liturgical Texts* (Philadelphia: University of Pennsylvania Press, 1917), 317-318.

[5] Barton, *The Royal Inscriptions of Sumer and Akkad*, 221.

[6] Ibid., 231.

[7] Martin A. Beek, *Atlas of Mesopotamia* (London: Thomas Nelson & Sons Ltd., 1962), 21.

[8] Andre Parrot, *The Tower of Babel* (New York: Philosophical Library, 1954), 26-27.

[9] Apparently the tower would be as high as the netherworld was low.

[10] James B. Pritchard, *Ancient Near Eastern Texts Relating to the Old Testament* (Columbia: University of South Carolina Press, 1955), 68-69.

[11] Parrot, *The Tower of Babel*, 18.

[12] Ibid., 19-22.

[13] Ibid., 22-23.

[14] Isaac Preston Cory, *Ancient Fragments of the Phoenician, Chaldean, Egyptian, Tyrian, Carthaginian, Indian, Persian, and Other Writers* (London: William Pickering, 1832), 52.

[15] Parrot, *The Tower of Babel*, 40-41.

[16] Ibid., 48.

[17] Hawkes, *The Atlas of Early Man*, 46-48.

[18] Parrot, *The Tower of Babel*, 64.

[19] *Webster's New Collegiate Dictionary*, 1979.

[20] Parrot, *The Tower of Babel*, 64.

[21] Hawkes, *The Atlas of Early Man*, 92.

[22] Lamberg-Karlovsky and Sabloff, *Ancient Civilizations: The Near East and Mesoamerica*, 163.

[23] Ibid., 165.

[24] Edmond Sollberger, *The Babylonian Legend of the Flood* (London: The Trustees of the British Museum, 1962), 12.

[25] Lamberg-Karlovsky and Sabloff, *Ancient Civilizations: The Near East And Mesoamerica*, 165.

[26] Samuel Noah Kramer, *Lamentations Over the Destruction of Ur* (Chicago: University of Chicago Press, 1940), 39-41.

[27] Ibid., 45.

[28] Ibid., 45.

[29] Lamberg-Karlovsky and Sabloff, *Ancient Civilizations: The Near East and Mesoamerica*, 167.

[30] H. H. Rowley, *Atlas of Mesopotamia* (New York: Thomas Nelson, 1962), 45.

Chapter 18

CONCLUSION: NOT THE END, BUT A BEGINNING

When I was an undergraduate in college, my Freshman math instructor offered this gem of wisdom. He demonstrated an involved formula to solve a problem. One classmate asked if another method would work just as well. "Oh, you could use another method," the student was assured, "you just won't get the right answer." Sometimes knowing the right answer can help us select the right reasoning.

The right answers in the origins issue are not confined to God's Word in Scripture alone, but need further defining by studying God's works in nature. The reverse is also true. The book of works can be clarified by the Book of Words. The danger, of course, is we may let overemphasis on either one contaminate the evidence provided by the other. This has been a prime source of error causing needless conflict.

Continually the Bible proves to be trustworthy and dependable for the things intended. Paul delineates these in II Timothy 3:16, 17: "All Scripture is given by inspiration of God, and is profitable for doctrine, for reproof, for correction, for instruction in righteousness: that the man of God may be perfect, thoroughly furnished unto all good works."

Notice that the apostle did not say the Bible was suitable for beating scientists over the head with it. Nor should men of science ignore the ample documentation the Bible itself imparts, authenticating its own authority. The Bible does speak to science in places; however, it was not intended to replace science text books.

Science primarily is a process with guidelines, a means to obtain solutions, a methodology for arriving at the truth. Christians need not fear the truth, but should welcome the truth no matter who finds it.

The conclusions presented in these last pages are based upon an honest examination of pertinent evidence, both Scriptural and natural. There will still be unanswered questions, but hopefully, the thorniest difficulties have been dealt with herein and put to rest.

Ancient Ancestors

One aspect of the evolutionary history of man's origins will probably be resolved in the next few years. The human genome mapping project is cranking out revelations almost on a weekly basis. Faulty genes are being identified and linked to corresponding illnesses. In their search for data, DNA molecules from lower creatures are studied continually as stepping stones toward solving the human DNA puzzle. A likely result of this research will be the proof concerning man's origins as it pertains to non-human predecessors.

Proof of common ancestry, if it comes, would not prove further aspects of evolutionary theory such as a naturalistic explanation for the Big Bang or the origins of life, but it would certainly give atheistic evolutionism a shot in the arm. Many would conclude that if one key feature of the overall theory is confirmed, then the entire theory of evolution would be proven. Such a conclusion would be unfounded, but it would certainly gain followers.

From whom or what did prehistoric man derive? Did man evolve from non-human primate ancestors? At this point in human knowledge, we can choose for ourselves and cling to hope. If conclusive data comes in confirming shared common ancestry, however, it will be time to abandon a sinking vessel, one that is quite waterlogged already.

Whatever answer escapes the realm of theory and attains the status of established scientific fact, the result, shared common ancestry or separate acts of creation, will not detract in any way from God's truth or His creative power.

Although the idea may offend some innate sensibilities, we cannot refute nonhuman ancestors based upon the biblical narrative. The fact of the matter is late in coming, but nevertheless, the Bible does not fall on either side of that divisive issue. Scripture neither confirms or denies a genetic link between man and lower forms of animal life.

What the Bible is silent about should not overshadow the things it maintains. Each human being has access to God's promise of salvation by virtue of the sacrifice at the cross. It matters not whether we came from Adam or mitochondrial Eve. Just as sacrifices of unblemished animals are no longer required, neither are bloodlines a factor in obtaining God's kingdom.

> Redemption is no more dependant upon the lineal descent of all mankind from Adam, than it is dependent upon their lineal descent from Abraham, the father of the faithful. [1]

Dissent

Traditional apologists invoke a handful of Bible verses in order to support the traditional claim that all of humankind must somehow be Adam's descendants. We will look at these to see whether their contentions are on solid or shaky ground.

The First Man, Adam

I Corinthians insists, some contend, that Adam must be the ultimate biological progenitor of all mankind. "The first man Adam was made a living soul; the last Adam [Christ] was made a quickening spirit" (I Cor. 15:45). After examining this and verse 47, M'Causland concluded:

> These passages cannot be construed as declaratory that Adam was the first created human being, either directly or by implication. He is designated the first Adam or man, only as contradistinguished from the "last Adam,"

or "the second man," and not in relation to the other items of the human family. Were they to be construed otherwise, the necessary results would be, that Adam was the first, and Christ the second, created man, or that Adam was the first, and Christ the last, of the human race--neither of which propositions are true. [2]

There are at least four manuscripts, including the Vatican's text, where the word "man" is not found. [3] These may be read: "The first Adam was made a living soul: the last Adam, a quickening spirit ..." Just as Christ in no way was the last man in any biological sense, the parallel would seem to indicate that Adam, likewise, was not the first. A literal reading of the text may prohibit rather than permit Adam from being the first human.

The Mother of All Living

Adam named his wife Eve "because she was the mother of all living" (Gen. 3:20). Does this verse signify that all present-day human populations are derived from Adam? It would if the word "living" was synonymous with *Homo sapiens*. But "living" is an adjective, unless it stands in for a noun, such as "life." Was Eve the mother of all *life*? The word "life" surely carves out far more territory than the verse intends.

If "living" is an adjective in this verse, it modifies a missing noun. We could pencil in "men" for the missing word, but this is little different from the phrase "the first man Adam" (I Cor. 15:45). Again, we have the same definition problem addressed in chapter 10. Should "man" be defined broadly or narrowly?

In this verse, the Hebrew word *chay*, translated as "living," can be either an adjective or a noun. Used as a noun, it can simply mean "relatives." In this sense, all of Eve's relatives (Adamites) emanate from her. By inference, Non-Adamites can look elsewhere. The real significance of Genesis 3:20 is that it establishes an acceptable standard for marriage. By this verse, Adam had no other wives or concubines.

Delaware license plates brandish the phrase, "The First State." The Delaware legislature was first to endorse the Declaration of Independence and declare itself independent from England. The qualifier, "in the U.S.," is unstated, so to speak. Certainly, Delaware claims no world ranking. The same is true of Adam. He was first in the Old Testament, and if Alulim is the Sumerian equivalent of Adam, then he was the first man in recorded history as well.

Why get bogged down in semantics? What does "man" mean? Does "man" identify humanoids, hominids, all in the genus *Homo*, archaic *Homo sapiens*, modern *Homo sapiens*, Caucasians, or Adamites-Semites-Israelites-Jews? Picking *Homo sapiens* out of the lineup is just as arbitrary as picking any one of the other categories of precursors.

Notice that Jesus avoided any semantics problem. After the resurrection, the new covenant encompassed both gentiles and Jews. In Mark 16:15, the risen Christ commissioned His disciples to "preach the gospel to every creature." Any less encompassing terminology might have left them to wonder who was eligible. The word "creature" leaves no latitude for exclusion. Everyone, regardless of race, color, gender, or national origin is deemed suitable for God's kingdom.

Quantifying the World

In addition to the Genesis account of the flood, the Apostle Peter makes mention of the flood in his second epistle. "Whereby the world that then was, being overflowed with water, perished" (II Pet. 3:6). Does the word "world" in this verse identify Noah's flood as a global deluge? Not necessarily; look at how we use "world" in English.

We say "world" every day without intending it to mean the entire continental land mass of our planet. The wide "world" of sports may "span the globe," but only covers sporting events. The "world" of entertainment concerns films, TV, and stage plays. The "world" of fashion includes cosmetics, hair styles, and clothing. So although we sometimes mean planet when we say world, we may also use "world" to connote the encompassing of all of a category of something. Bible writers did the same thing.

"World" was sometimes used in the sense of planet (what little they knew of it). In John 21:25, "I suppose the world itself could not contain the books that should be written." Many times, however, the Bible authors used "world" to mean just the human occupants. "But the world may know that I love the father ..." (John 14:31). Also in John 3:16, "For God so loved the world ..." applies to human beings living in the world.

The word "world" also denotes only a segment of humanity. James teaches, "the friendship of the world is enmity with God" (Jas. 4:4). Here James grouped the unsaved, or unrighteous, into a "world." Doing just the opposite, John used "world" for the redeemed. "Behold the Lamb of God, which taketh away the sin of the world" (John 1:29).

Peter himself narrowed the "world" by speaking of the flood being wrought upon "the world of the ungodly" (II Pet. 2:5). The "world" that was "overflowed with water" and "perished" singles out the Adamite world including some hapless bystanders. Nobody headed for high ground in the African world though, or in the Asian world, or in the Americas. Even the nearby Egyptians, Persians, and "Nephilim" too, for that matter, were survivors untouched by the Genesis deluge.

A Local, Recent Flood

The duration of Noah's flood, taking over a year from beginning to end, may seem to be an effective counterpoint to a purely Mesopotamian episode. The flood's duration does not prescribe the scope of the flood though there are feasible, alternate explanations.

A sequence of two or more flooding episodes may have stretched over two rainy seasons. A violent typhoon may have deluged the Mesopotamian valley, damming up the Euphrates river for a time, taking some months for the water to drain away. God's hand of judgment may have stretched out in a miraculous intervention similar to the destruction of Sodom and Gomorrah and the parting of the Red Sea. Whatever happened, we need not lapse into a state of confusion simply because we do not know exactly what transpired.

The date of the flood also is an issue we can debate into the next millennium. Archaeological evidence places the flood at about 2900 BC. Traditionally, conservative Bible scholars set the flood at about 2348 BC. Even with the later date, there is a meager number of generations between Abraham and David to span over a thousand years. An even earlier date for the flood would seem to exacerbate that problem further.

One possible solution is that some generations could be missing from the text. A few critics have pointed out that the "fourteen generations" between David and the "deportation to Babylon" (Matt. 1:17) should have been seventeen generations because Ahaziah, Joash, and Amaziah are found in II Kings 8:25-15:1, but are missing in Matthew. Postulating extra generations is one conceivable method of covering the great number of years between Abraham and David, but there is another possibility.

New Testament authors drew on the Septuagint in circulation 200 years before Christ. The Greek text puts 1,072 years between the flood and Abraham. Subtracting 1,072 years from the archaeological date of 2900 BC for the flood, and taking David's birth at a little before 1030 BC, puts less than 800 years between Abraham and David, which is more easily covered by fourteen (or seventeen) generations than the over 1,000 years conservative scholars have now. There is some consensus among Bible scholars, however, that places Abraham at around 2055 BC. If we work back from this date using the Septuagint's figures, it puts the end of the flood at 3127 BC, not an unreasonable timeframe. Thus, there are at least two ways the archaeological date for the flood can be closely reconciled with a date derived from the biblical genealogies.

To be sure, there will be a measure of dissatisfaction with a local flood in the neighborhood of 2900 BC, impacting only a fraction of the human race. Some would prefer a universal flood in the distant past, or if it was local, it had to have happened at the predawn of history to ensure Adam at the apex of humanity and Noah as the second founding father. Others will cling to a global flood at 2348 BC to concord with the Ussher chronology.

What these "die hards" must do if they wish to preserve inde-
fensible beliefs is (1) find a way to negate the confirming data and
historical documentation for a relatively recent episode in Southern Mesopotamia, and (2) argue for a mankind obliterating flood
either early or late in the absence of supporting biblical or natural
evidence. And what is to be gained? Just the continued
back-of-the-hand treatment from scientists and academics they have
endured up until now.

"Whoever shall smite thee on thy right cheek," Jesus said, "turn
to him the other also" (Matt. 5:39). To be struck on the right cheek
by a right hand is to be hit with the back of the hand. It is degrad-
ing, an insult. By turning the other cheek, we are offering ourselves to the forehand; it's still a beating, but not belittling. We
may take a thrashing for the gospel's sake, but we need not bear
public ridicule unnecessarily for what we believe and profess.

Revising the Text

Although history and science should not sit in judgment of
Scripture, historical and scientific data can provide a background
of information from which we can make necessary adjustments in
our interpretations. Though God could have brought all this about
any way He desired, even He has a self-imposed restriction. God
cannot lie.

Bible publishers and expositors have a serious obligation to
handle God's Word so as not to cause our brothers to stumble. Some
hard looks need to be taken at confusing terminology with a view
to resolving difficulties, not promulgating them. The unfortunate
tendency has been to canonize and cherish as unalterable the English language text of 1611 AD. Those 17th century translators were
as hindered by ignorance as present-day translators are bound by
tradition. Ignorance and intransigence serve to inhibit our unsaved
brothers from accepting the Christian message, and this can be rem-
edied only when publishers of Bibles and commentaries demonstrate a passion for the truth that surpasses their desire to avoid
controversy. But revision must come. How can pastors "rightly
divide the word" when it has been translated incorrectly?

The idea of using "slime" for mortar (Gen. 11:3) would run shivers down a bricklayer's spine. An inspection of a temple ruin in present-day Iraq would reveal that naturally occurring asphalt from nearby tar pits was used as a weather preservative and adhering agent. This is a prime example where God's general revelation should be used to define special revelation. Who cares what the translators thought almost 400 years ago? Change the text!

"Sky" is more accurate than "heaven" in Genesis 1:8. "Whales" and "fowls" (Gen. 1:20, 21) are as inappropriate as sea serpents and dodos. "Stream" should supplant "mist" in Genesis 2:6. "Earth" should give way to "land" in the same verse. And while the broom is out, the word "earth" should be swept from Genesis in all but a handful of instances to alleviate needless ambiguity. We need to make "hills" out of "mountains" in Genesis 7:20 and 8:4, 5.

In 1696, William Whiston recognized the contradiction of the mention of daylight on the first day of creation with the sun, moon, and stars functioning as celestial clocks on the fourth day. He recommended using an optional tense for the verb in Genesis 1:16 for clarification.

Whiston's version was "God 'had' made the two great lights," and " 'had' made the stars," to connote that these events were finished in the past, and to shift focus away from the making of the luminaries, and toward their function as timekeepers.

There is a lack of consistency when "man" and "Adam" appear in Scripture, and the translation often deviates arbitrarily from the original Hebrew. Although 'adam sometimes is translated "Adam" in the Old Testament, it also appears as "man." "Man" or "a man" appears in Genesis 1:26-27; 2:5,7. "The man" or "Adam" is found in Genesis 2:8,15, 16, 18, 19, 20, 21. Adam is commingled with the word "man" in Genesis 2:21-23, where God "caused a deep sleep to fall upon Adam," and took the rib "from man." This haphazard use of "man" and "Adam" causes confusion. Of course, up until the 19th century, man and Adam were thought to be rather synonymous.

Here, as usual, the Bible itself can guide us through this difficulty. Jacob became "Israel," a name bestowed upon him by his

night visitor (Gen. 32:24-28). Where the Bible mentions Israel
during his lifetime, it refers to the man previously called Jacob.
After his death, "Israel" denotes the nation of Israel consisting of
primarily, though not entirely, the descendants of Jacob.

Adam (the Hebrew *'adam*) is first applied to the man created
in the image of God and placed in the garden. This usage should
be retained in translations. For example, "God created Adam [not
"man"] in his own image ..." (Gen. 1:27). "This is the book of the
generations of Adam. In the day that God created Adam [not "man"]
in the likeness of God made He him" (Gen. 5:1). After Adam's
death, starting with Genesis 6, a variation should be used such as
"son of Adam," "descendants of Adam," "Adamite," or "Adamites."
There may be room for an occasional exception where an exten-
sion of meaning may be applied to all of humankind. For example,
Romans 1:16 uses "Greek" as a collective for all who were not
Jews.

Remember, the Old Testament was written in Hebrew and Aramaic, whereas New Testament authors wrote in Greek. The phrase
"son of man" appears in both the Old and New Testaments. It can
be perplexing when *'adam* is translated as "man," especially when
a word for generic man or mankind was used in the original, *'ish* in
Hebrew or *'enash* in Aramaic. Psalm 8:4 is a case in point: "What
is man, that thou art mindful of him and the son of man, that thou
visitest him?" How are men different from sons of men? Why the
redundancy? Aren't all men sons of men? Yes, but not all men are
sons of Adam! In this verse, God is "mindful" of "man" (*'ish*), but
it is the sons of Adam (*'adam*) that He "visitest."

In Psalm 80:17, "Let thy hand be upon the man [*'ish*] of thy
right hand, upon the son of Adam [it should read] whom thou madest
strong for thyself." See what a clarification it makes in Numbers
23:19: "God is not a man [generic man], that He should lie; neither
the son of man [Adam!], that He should repent ...."

The prophet Jeremiah harkens to the destruction of Sodom and
Gomorrah. By way of translation, in one sentence he appears to
say the same thing twice: "... so shall no man abide there, neither
doth any son of man dwell therein" (Jer. 50:40). Here again *'ish*
and *'adam* are both translated "man." Had the translators let *'adam*

be "Adam" instead of "man" we would know that neither Adamites nor Non-Adamites live there.

God repeatedly calls Ezekiel the son of Adam (*'adam*) starting in Ezekiel 2:1, although translators render him "son of man" thus blurring the distinction between the prophet and Christ, a distinction delineated in the book of Daniel. Pointing to the coming Messiah, Daniel relates a vision: "... and, behold, one like the Son of man [*'enash*] came with the clouds of heaven ..." (Dan. 7:13). Yet Daniel is addressed: "Understand O son of man [*'adam*]: for at the time of the end shall be the vision" (Dan. 8:17). If only the vision of the translators had been equal to the vision of Daniel.

In the New Testament, occasionally Christ is called "son of David," but more often, "son of man." All four of the gospels include this phrase repeatedly respecting Christ. We also find "son of man" in Acts 7:56, Heb. 2:6, and Rev. 1:13; 14:14. In every instance "man" is the Greek *anthropos* meaning "human." The phrase "son of man" should be reserved for Christ who is nowhere called the son of Adam. To differentiate, a prophet should be called "son of Adam," not son of man.

New Translations - Same Problems

One might think that newer translations of the Bible would bask in the light of recent scientific discovery and accord more closely with history and anthropology. The New Jerusalem Bible (copyright 1985) was praised as: "truly magnificent" (*Journal of Biblical Literature*); "the best of modern scholarship ...," (*Christianity Today*); and "the best of modern translations" (*The Living Church*). 4 And, sure enough, where the KJV slips in "slime" in Genesis 11:3, the NJB sticks with "bitumen." "Whales," "grass," and "fowls" in the KJV are replaced by "sea-monsters," "vegetation," and "birds" respectively in the NJB. Yet, in instances, this latest tribute to modern scholarship proves more troublesome than the antiquated King James Version.

A subtitle for the first chapter of Genesis in the NJB reads as follows: "The Origin of the World and the Human Race." Already the translators have ventured an opinion before the Bible has a chance to say, "In the beginning ..."

The "kinds" of animals God created (Gen. 1:25) has been superseded by the word "species" in the New Jerusalem Bible. Why not "genus" or "phyla" perhaps? All of the present-day phyla can be traced to the Cambrian period. An argument that God specially created all the existing phyla, including some that are now extinct, at or just prior to the Cambrian period at least could be defended rationally without undue contradiction from the fossil record. But to speculate that an act of special creation was required for each and every species defies the fossil record and the genetic evidence. Use of the word "species" rules out a common ancestor for the horse and zebra, for example, something that would get a howl out of biologists. At least the word "kinds" was open to interpretation, whereas "species" is a loaded term.

Jubal was the "father" of those who play the harp and organ (Gen. 4:21) in the KJV. He is their "ancestor" in the NJB, a less elastic designation. Addressing the scribes and pharisees, Jesus told them their "father" was the devil (John 8:44). It is unlikely the human objects of His scorn were direct-line descendants of an evil spirit being, just as it unlikely that all harpists and organists descended from Jabal. In Genesis 4:22, Tubal-cain was an "instructor" of those who worked with brass and iron in the KJV. And like Jubal, he becomes their "ancestor" in the NJB. Were all metal workers the offspring of Tubal-cain?

It has been shown that the flood account in the KJV can be local versus global once we account for Hebrew manners of speech, verbiage, and syntax; the use of "perfect" numbers; alternate meanings of key words, etc. The men destroyed in the flood can easily be those of the covenant, the descendants of Adam. This accommodation is not permitted in the New Jerusalem Bible, "... I shall rid the surface of the earth of the *human beings* whom I created ..." (Gen. 6:7).

The "fountains of the deep" (Gen. 8:2) in the KJV clearly refers to an irrigation system, and the Atrahasis epic contains the identical phrase. This Assyrian legend of the flood not only defines the term by usage, it demonstrates a commonality with

Genesis. The translators of the NJB decided that "springs of the deep" was a good choice of words, thereby destroying the meaning and disguising the commonality.

In the KJV, Nimrod was "a mighty one in the earth" (Gen. 10:8), leaving room to negotiate his exact chronological ranking, whereas the NJB designates him "the first potentate on earth." Narmer, the first pharaoh of Egypt, predates this period, and certainly so do the pre-flood rulers in Southern Mesopotamia.

In short, newer Bible versions such as the New Jerusalem Bible do not alleviate difficulties in Genesis, and may actually exacerbate them. Modern translators do not acknowledge scientific evidence, they ignore it. The same mental block that hindered translators in the 1600's also hamstrings modern translators, namely, presumptive bias. In the case of the New Jerusalem Bible, it meanders into murky waters through typical young-earth creationist lingo, whereas the old King James Version steers a more neutral course.

Necessary updating is long overdue, but even newer translations will not ease confusion until translators take extra-biblical evidence into account. This is not to imply that a wholesale disruptive onslaught should be unleashed on the inspired text; however, English equivalents should be re-examined in light of alternate meanings from the original Hebrew. Words from variant texts should be considered in some instances. The goal should be to eliminate conspicuous mistakes. If we lose the reader in the first eleven chapters of Genesis, he may never make it to the twelfth, let alone to the New Testament.

After removing obstacles in the text, an enlightened interpretation in accord with present-day knowledge can follow. A literal interpretation of Genesis, in light of modern discovery, simply works best in harmonizing the Genesis account with the evidence of the natural processes God invoked to affect His creation.

Creavolution

The Hebrew word *bara'*, "rendered "created," used seven times in the first two chapters of Genesis, is connected to three definite acts: the creation of heaven and earth, the creation of sea life, and the creation of a specific man and a specific woman. The man God created was Adam, and the woman was from Adam (Gen. 5:1-2; I Tim. 2:13).

Clearly, the word "created," where it pertains to man, identifies Adam (and Eve). Notwithstanding, many Bible expositors employ an act of special creation for generic man defined as *Homo sapiens*; and some would include extinct hominids too. This is an unwarranted extrapolation that follows from misreading the Genesis text.

The word "create" best describes God's bringing into existence the universe (Gen. 1:1), primitive life (Gen. 1:21), and Adam and Eve (Gen. 1:27). In other instances, God "made," which is the Hebrew word '*asah*, or He "let," or He "let the earth bring forth." What we find is that Scripture stipulates acts of special creation at only three junctures, and is unspecific when another process may have been in operation, evolution, for example.

The term "creation" can cover all the billions of years since the inception of the universe. The entire event from the beginning may come under the blanket of creation. When we are speaking of a process of change through time, however, the appropriate word is "evolution." The problem is that the term "evolution" has been tainted by abuse. It has been sullied by exploitation and rendered unserviceable to describe a process ordained by God. Theistic evolution has its own agenda, and is similarly unsuitable.

"Creavolution" is a term that can describe God's appointed process whereby laws of nature are permitted to function, and organisms are allowed to adapt to their environment with increasing complexity without direct divine intervention. If creation is the factory, creavolution is the machinery. Essentially, creavolution is a "hands off" process except for the three punctuations outlined in the Genesis test, the creation of the universe, primitive life, and Adam and Eve. Those who call for additional acts of special creation proceed without biblical support.

Following the first punctuation, the Big Bang, natural processes adhering to prescribed laws of physics and chemistry operated for billions of years. After God's creation of primitive life, the second punctuation, natural life processes, coupled with an inherent adaptive capacity, produced the earth's abundant and varied animal life. In time, this process alone could have yielded up *Homo sapiens* even though there are many who would prefer to believe that we have no roots to anything that swung from trees. Adam and Eve were specially created, and comprise the last punctuation.

Creavolution as a process can endure "random changes" in the nucleotide bases of DNA molecules in living creatures occurring through time, such that no unseen hands needed to manipulate the process. "Random" is not to be taken as synonymous with "accidental" or "unforeseen" inasmuch as the end result is known to God. If we wish to establish that the Creator does not preempt or force His own choices continually upon nature, a conclusion that seems altogether valid, then the term "natural selection" suffices, and describes an unorchestrated process. Thus "random changes" and "natural selection" are entirely acceptable nomenclature within the creavolution process.

The mechanism for change preferred by biologists so far has been built on the premise that genetic variability exists naturally in living organisms. Changes in DNA arise from chance mutations, "chance" to include meaning "unforeseen." This is an unsubstantiated premise endorsed and promulgated by the scientific community. The data suggests that a mechanism not yet understood was installed in living organisms to give them adaptability. This grooming of the DNA, benefiting organisms with inheritable modifications, can be inferred from nature's evidence. Further exploration in microbiology may give us a better understanding eventually.

Only a few dissenting voices have been raised so far to contend that the environment can impact gene mutations directly in some way, and thereby change the structure of subsequent generations. Hopefully, more will come. What we need to understand is that in this process the Creator is never caught unawares by the outcome. What has been known to God "before the foundation of the world" cannot be unforeseen.

Imago Dei

Bible expositors have taken the phrase "in the image of God" and blown it into proportions far beyond the simpler intentions of the text. An "image" is a likeness or representation of something. In Leviticus 26:1, the children of Israel were told to make "no idols nor graven image." Idols themselves can become objects of worship, obscuring the one true God who accepts worship directly. "The image of Baal" (II Kings 3:2) was an object of pagan worship, being a representation of that false deity.

In Genesis 1:27, Adam represented God, having been "created in His own image." This status was passed through the godly line of Seth (Gen. 5:3). Noah and his generations were God's chosen people, and thus were "in the image" (Gen. 9:6). This status as representatives of God was conferred upon the Israelites through the Abrahamic covenant (Gen. 17:1-8).

Apparently, those outside the nation of Israel were outside the realm of accountability. This can be inferred from Matthew 23:15, "Woe unto you scribes and Pharisees, hypocrites! for ye compass sea and land to make one proselyte, and when he is made, ye make him twofold more the child of hell than yourselves."

When one outside the Jewish faith was brought to the knowledge of God, he became accountable. Because of false teaching, he was condemned. This unique status for Israel as God's chosen people was rescinded, or at least modified, at the cross. Christ was appointed by God as His representative. The second Adam, Christ, was in the "image of God" (II Cor. 4:4) just as the first Adam, and the mantle was passed to the followers of Christ.

In I Corinthians 11:7, Paul's instructions were not to unregenerate men, but to the redeemed of the church at Corinth. According to Paul, they were in "the image and glory of God." They received this authority as believers in Christ, "who is the image of the invisible God, the firstborn of every creature" (Col. 1:15). Fallen man has no claim to God's image unless he receives it through redemption.

Psalm 8 points to the coming Messiah. David affirms that Christ has dominion over all things. This was given to Adam at his creation (Gen. 1:28), and was intended for his generations, but it was clearly in Christ's hands after the Fall. "Thou madest Him to have dominion over the works of thy hands: thou hast put all things under His feet: all sheep and oxen, yea, and the beasts of the field; the fowl of the air, and the fish of the sea ..." (Psa. 8:6-8).

Dominion over the lesser animals docs not accrue to man. It was inherent first in Adam, and then in Christ. Those who belong to Christ share in His authority and in His dominion. Those who are not in Christ, though they may act as if they have divine permission, merely usurp an authority not granted by God.

The notion that all of mankind has "dominion" over the earth and were created in God's "image" derives from the mistaken idea that Adam was the ultimate progenitor of the human race. From this, Bible expositors have gone overboard postulating the marvelous similarities between us and our Creator.

In what manner are we, his stumbling creatures, like the Most High God? Do we possess His holiness, or His righteousness? Can we boast of His wisdom? Are we omnipotent? Can we transcend time? Is it in our power to forgive sin? Can we grant immortality? No, we mere mortals presume too much.

Our claim to being in His image is on the righteousness of Christ, not by any birth right, lest any man should boast. "For my thoughts are not your thoughts, neither are your ways my ways, saith the Lord" (Isa. 55:8).

The Human Soul

If we consider the possibility that human beings evolved, when or how did we become endowed with souls? A certain unease can come with the idea of sharing common ancestors with creatures more hairy. Most would prefer to think that man was specially created, thus barring descent from any oddball ancestors. It fits our prideful nature. It might do us good to ponder the likelihood of humbler beginnings.

If we had a choice, we might think it is more desirable to be direct descendants from Adam and Eve, who were created without ancestors. We could say that because of our unique origin, we had "souls" (some prefer "spirit"), and all those lesser quality evolved creatures like *Homo habilis*, *Homo erectus*, and the curious Neanderthals were used as models perhaps, or experimental prototypes.

How do we get those wonderfully unique and redeemable (or punishable) souls? Are we endowed with souls because of direct descent from specially created ancestors? That has been the traditionally espoused rationale, but in essence, it works no better than gap theory.

All living mammals, including man, carry the potential for life to bring about further offspring. Until union takes place between an egg and a sperm, there is no physical being capable of further replication. Life, as evidenced by cell division, commences at conception.

If a "soul" came from the man, we would have to postulate little potential souls as an integral part of each one of countless billions of sperm cells which do not exist in any form until the human male reaches puberty. After that, these are produced throughout a man's entire life. Would human males manufacture souls continually to accompany tiny sperm cells? That looks doubtful, so how about the woman?

Women are supplied with hundreds of eggs at birth. After puberty, a woman releases an egg approximately every 28 days until she reaches menopause. Do those eggs come prepackaged with souls too? If so, far more end up in waste treatment plants than in choir lofts. That is not God's plan. We must differentiate between the material and the spiritual.

Our DNA, and all that is associated with our life-producing capabilities, are simply physical processes for producing physical beings. It is hard to imagine any spirit entity intertwined with the potentials of life. When union takes place, that is a different story. But how spiritual souls match up with physical bodies is anybody's guess. Are souls allotted at conception; handed out at "viability"; assigned at birth; imparted at an arbitrary age of reason; bestowed at an arbitrary age of accountability; conferred at the acceptance of Christ's Lordship; or none of the above? If anyone knows the answer, please step forward.

The spiritual world is almost entirely out of bounds for human investigation. The angelic hosts, who exist without physical bodies, are living entities in a real sense just as we are. They think, interact with humans as "ministering spirits," have free will, and so forth.

Whatever part of us belongs to the spirit realm may be wrapped up in our material selves to some degree, but clearly we exist apart from our physical bodies, inasmuch as we were known before the "foundations of the earth," and glorified bodies will be conferred eventually.

We can gain some insight from Jacob and Esau (Gen. 25:20-34). In a sense, Esau was condemned from the beginning because his life's deeds were already pre-known to an omniscient God. Souls capable of redemption were part of the make-up of both brothers. Both were in the Adamic-Abrahamic line. Yet Esau's celebrated ancestry offered no guarantee of redemption for his soul. This suggests that redemption and Adamic ancestry are two separate issues, and are not interdependent.

If there is no commonly accepted theory as to how humans would get souls, even for direct-line descendants from a specially created forefather who presumably had one - Adam, not knowing how humans with evolved ancestry would be imbued with souls (should evolved ancestry be confirmed) is no worse case. The prospect of nonhuman ancestors complicates no further what is already a perplexing issue.

It makes no difference how our predecessors got here. The subject of souls is perplexing regardless of our ultimate origins; however, this should not be a source of consternation. John 3:15 assures, "That whosoever believeth in Him should not perish ..." A soul or spirit capable of redemption is expressly implied in that "whosoever." The Bible does not say whosoever believeth, and has Adamic ancestry, should not perish.

Biblical Bloodlines

The Bible is not about human history. It is about Adamite-Semite-Israelite-Jewish history. It is about a particular bloodline that weaved its way through the past to culminate in Jesus. The early passages of Genesis demonstrate how strictly this plan of patriarchal history predominated. But consider this: Sargon of Agade was a legend in ancient pre-history. From cupbearer to the king of Kish, by bold conquest, Sargon ruled over all of Sumer and Accad. Sargon began the Sargonid dynasty, changing the recorded language of the region. Cuneiform texts on clay tablets display a dramatic shift in the official tongue from Sumerian to Semite with Sargon's reign. Yet, Sargon rates a zero in the biblical text. This is paramount to omitting Einstein from a list of famous scientists, or leaving out Napoleon in a catalog of French emperors.

The king of Assyria (Isa. 20:1) was named after the great Sargon, proving the ancient ruler was not forgotten. History books have designated this Assyrian king as Sargon II to avoid confusion. Why is the name of the great Semite ruler absent from the Genesis text? Simply, the Genesis account is not about important people, but important bloodlines. Sargon had a dubious beginning. His father was untraceable, rendering the fabled king unsuitable for inclusion. In this regard, Sargon was an outsider, being outside the line of promise leading to the Messiah.

For those of us who are outside the covenant bloodline from Adam to Abraham, we still can be note-taking observers. We can anguish over the trials of the Jews at the hands of their brutal captors, feel their passion for their God, wonder at their continued failings, and admire their unfettered zeal. We can sense their innate stubbornness, and we can applaud their uncompromising willingness to set down a true history regardless of how embarrassing or self-effacing that may have been.

Most of us are outside their bloodlines and not part of Semitic history, but we can borrow a little. More importantly, we can learn of the character and nature of their God. We can see how He dealt with their disobedience. We can esteem God's holiness, search His truth, and dwell on His righteousness.

We can gain a sense of how He loved His chosen people, how He called to them continually, and forgave them, and remained faithful to them. And above all, we of no noble parentage can come boldly before the throne of their God, and kneel before Him, and claim their gift. We can be adopted into His family.

> The penalty of death has been abolished, and the for-
> feited inheritance redeemed for Adam's race; and all other
> races are admitted to the participation of the benefits pur-
> chased by the Saviour's blood. [5]

A Summation

The universe began. God takes credit for that at the outset, in the first sentence of Genesis, to quell all doubt. No credible theory has been able to dislodge a creation event as the first cause. Our Creator, the Prime Mover, also established laws of physics and chemistry that regulated His creation to culminate in our particular solar system featuring space station Earth.

Life with its beauty, organization, and complexity owes its start to the Creator. Let biologists question that prospect to their heart's content if they have nothing better to do. Even if scientists were to synthesize a primitive replicator in a laboratory, all that would prove is that man can play God now and again, and may get somewhat similar results.

By studying God's ways and applying our own intelligence, we may someday be able to duplicate some aspects of God's cre-ative endeavors. What would that prove? We did not unravel the complexities of nuclear power, and deduce that the fission process was an accident of nature.

If ever we do produce life through a heavy application of our own mental prowess, how could we conclude that the beginnings of life on earth were only lucky, chance happenings? We cannot use our own intelligence to produce something like life, and then declare that creating life required no intelligence. God told us He created original life; why not trust Him?

Man is the final product, and has a fossil history that suggests his coming into pre-eminence gradually. We have no more reason to fear the evidence of man's ancient beginnings than we have to fear any other facets of God's general revelation.

> There is no more difficulty in realizing the existence
> of a world of human beings without any knowledge of
> God before the creation of Adam, than there is in con-
> ceiving that millions of human beings have been ever
> since, and are, at this moment, living wholly ignorant of
> the name and nature of God, in a world on which the Sun
> of Righteousness has shined. 6

Not the first mammalian, bipedal couple, but the first covenant couple, Adam and Eve, were specially created about 5000 BC. In Genesis 1:27, the Hebrew word *bara* for "create" appears three times in this verse for emphasis. Adam's late insertion into humanity precludes his being ancestral to all present-day populations, even though we are all related to one another. Apparently, all human beings sprang from one common genotype about 200,000 to 100,000 years ago.

The evidence of nature and archaeology certifies the flood of Noah's day as a deluge narrowly confined. The indigenous Sumerian populations proved to be no match for Noah's emerging progeny who usurped not only their women, lands, customs, and method of writing, but their polytheistic religion as well. The marauding Semitic peoples continued to flex their muscles over a broader area, impacting first Persia and then Egypt.

A select segment of Semites, through the line of Noah's grandson, Arphaxad, tried their hand at ziggurat building at Babel to their grief. It was through this line that God's faithful servant, Abraham, would emerge. In the midst of turmoil, God preserved a faithful remnant for His eventual plan of salvation.

A Change in Attitude

The greatest obstacle to understanding the issue of origins among believers has come from a propensity to assume that all of humanity somehow originated with Adam, survived a worldwide flood, and had their languages transformed at Babel. To a degree the Bible will tolerate that point of view if we insist on it, but the inspired text is not a fomenter of such an unworkable proposition, and as we have seen, it even negates those premises.

The largest impediment to understanding in the scientific community stems from an unwillingness to recognize that immutable physical and natural laws, even "fundamental constants" of nature, have been established by an extremely intelligent Law Giver. Seeking and identifying natural causative factors in life processes is not a heresy. But when a created environment and created life forms stand out as the best explanation for what we see in nature, then the scientific establishment should be willing to acquiesce.

Clearly, some soul-searching from both sides would be helpful. If we are ever to enjoy harmony, or at least a truce between the Christian and scientific communities, there needs to be a little give from both sides. In the interest of balance, here are six key areas where accommodation, maybe even agreement, should be possible.

The academic and scientific communities could acknowledge those areas where a Creator better fits the data than any other explanation:

1. At the beginning. The singularity where it all began, either from a tiny grain of matter, or from nothing at all.

2. At the start of life. No workable theory has been devised where chemicals can come alive and reproduce.

3. At the direction of life. The theory of evolution does not explain satisfactorily how life forms can achieve greater complexity without at least one intelligence input somewhere in the process.

The Christian community also must come to grips with erroneous Bible interpretations that have plopped unsuspecting followers into a quagmire:

1. At Adam's introduction. There is no plausible explanation that would allow Adam of Genesis to be ancestral to the entire human race.

2. At the flood. A global cataclysm, obliterating all human and animal life, is thoroughly repudiated by historical and archaeological evidence.

3. At Babel. The diversity of present-day races and languages preclude a common starting point at one of the Mesopotamian ziggurats.

One last word of caution: the use of "proof texts" torn from context to support flimsy models of creation that do violence to God's record in nature is a dangerous game. Those who choose to play it need to consider the damage they can cause to Bible credibility as perceived by nonbelievers.

The Beginning

Where we are today is difficult to know. We see scientists seeking alternative solutions to avoid the incontestable conclusion that we live in a created environment. On the other hand, some who profess God are actively engaged in science-bashing to the overall detriment of the gospel message. And, while liberal theology is on the road to apostasy, conservative Christian thought is headed toward irrelevance. It is enough to make grown men cry. Hopefully, this book will serve as a bridge, open some eyes, and produce more light than heat.

The more startling irony in all this is that while there have been some residents in the Christian camp calling themselves creationists, and eschewing science, there are scientists today, in their search for truth wherever it may lead, who are discovering the God of creation.

The materialistic Big Bang theory has fallen on hard times. Even though the sequence of events and effects following an incomprehensible explosion are agreed upon generally, knowing what was happening prior to that cosmic convulsion is a different matter.

Ask any physicist, "What was going on 10 minutes before the Big Bang?" The intriguing answer seems to be: there was no 10 minutes, or any time at all prior to the Big Bang. Time itself had a beginning. And, as regards matter and energy, a tiny ball of incredibly condensed matter cannot sit complacently in space for eons, and suddenly explode. Time, space, matter, and energy: all are created entities.

One scientist, Dr. Robert Gange, discusses an implication generated by sheer physics in his book, *Origins and Destiny*. In place of a theory based on materialism, he argues for a nonmaterialistic beginning, a New Inflationary Theory, called in some circles the "Inflationary Big Bang." According to Gange:

> In the Big Bang theory, our hands were tied because we could not go back in time to the actual beginning. A small impenetrable interval of time, called a "Planck time" separated us from seeing the true beginning. But the New Inflationary Theory frees us from this limitation and gives us a picture of the universe from the moment it unfolded. Were we to condense its implications into one sentence, it would be this: The universe seems to have come into existence out of nothing. That's right; out of nothing. [7]

Who can create something out of nothing? God can! And so, we have come full circle.

The physicist Paul Davies suggests that "science offers a surer path than religion in the search for God." [8] There is little doubt that science can find God, the Law-Giver, but can it find God, the Giver of *life*? What can science tell us about forgiveness, or about salvation and how to attain it? A question with eternal consequences we absolutely must answer for ourselves, and for our children yet to come, is not how or when - but why and who? Just who is this God?

In establishing priorities, deciding the "God question" demands first priority. For if we choose to worship a creator formulated by scientists, or deduced by philosophers, or proclaimed by progressive theologians, then make no mistake, it had better be that god, and not the God of the Christian fundamentalists, who judges.

NOTES

[1] Dominick M'Causland, *Adam and the Adamite* (London: Richard Bentley, 1864), 289-290.

[2] Ibid., 292.

3 Edward William Lane, Reginald Stuart Poole, ed., *The Genesis of the Earth and of Man* (London: Williams and Norgate, 1860), 58.

4 From the back cover of *The New Jerusalem Bible* under the heading "Praise for The New Jerusalem Bible," (New York: Doubleday, 1985).

5 M'Causland, *Adam and the Adamite*, 289.

6 Ibid., 287.

7 Robert A. Gange, *Origins and Destiny* (Waco: Word Books, 1986), 19.

8 Paul Davies, *God and the New Physics* (New York: Simon & Schuster, 1983), 229.